AFTER THE COVID VACCINE

Futurist Scenarios of Risk and Opportunity After COVID-19 Vaccines are Widely Available

JASON SCHENKER

AFTER THE COVID VACCINE

Futurist Scenarios of Risk and Opportunity After COVID-19 Vaccines are Widely Available

BY JASON SCHENKER

ISBN: 978-1-946197-96-2 *Paperback*
978-1-946197-72-6 *Ebook*

PRESTIGE
PROFESSIONAL PUBLISHING

Dedicated to those preparing for the new, new normal.

CONTENTS

CONTENTS

PREFACE

WHAT COMES NEXT?

What comes next?

It is a simple question at most times. But 2020 has most certainly not been most times. With the onset of the COVID-19 pandemic in March 2020, I got to work on my futurist book, *The Future After COVID*, which was released on 1 April 2020. I focused on the unprecedented impacts on business, the economy, and society as well as what would come next.

Now, standing on the threshold of global vaccination against COVID-19, it is time again to revisit the question of what comes next — and to again offer futurist perspectives.

The big idea of this book is to explore how the most important changes of the past year — during the time of Corona — will be reconciled in a future that allows us to counter, by some degree and over some time, some of the most deleterious risks and dangers of COVID-19 at the population level.

As in *The Future After COVID*, this book examines potential long-term changes, challenges, and opportunities that the COVID-19 pandemic is likely to hold for over a dozen different critical fields and industries, including the future of work, education, healthcare, supply chain, and more.

As I predicted in *The Future After COVID*, the impact of COVID-19 is likely to cast a shadow — in both bad and good ways — across the years and decades ahead. It will impact how we work, where we live, and what various industries will look like in the future.

Some of the themes, topics, and content in this book draw on my other futurist books as well as on research, courses, and training materials from The Futurist Institute's Certified Futurist program.

Changing Dynamics and Risks

This book represents an attempt to integrate the quickly changing landscape as COVID vaccination is about to become a reality. This means that not only was this book developed rapidly, but some of the realities, potential future impacts, and relevance of topics may change swiftly as well.

This makes writing a futurist book at a time of uncertainty quite a risk. But I believed it was worth it when I engaged in a similar process of writing *The Future After COVID*. And I believe it is worth it to share these views with the world to help people prepare for 2021 — and the years beyond. Hopefully, our expectations in this book will be as close to the future reality as most of our expectations in *The Future After COVID* proved to be.

It is increasingly important for analysts, consultants, and leaders to have a futurist framework to consider the potential impacts in the face of high uncertainty. This was true at the time of the COVID-19 pandemic breakout, and it is true now with so much riding on expectations for widespread vaccination.

I have written this book to address the expectations I shared for the long term in *The Future After COVID* as well as to share how those expectations may or may not differ in the face of what has been the most rapid development of any vaccine in history.

Of course, I also address some of the uncertainties that remain — the risks and opportunities — across sectors, industries, and professions.

Acknowledgements
No book is done completely alone. There are editorial, file conversion, design, and project management parts of the project to get a book like this completed. And those tasks require a team.

Along those lines, I want to thank The Futurist Institute and Prestige Professional Publishing staff for making this book a reality. And I want to especially thank **Nawfal Patel**, who managed the production of *After the COVID Vaccine.*

Most importantly, I want to thank my family for supporting me in my education, career, entrepreneurship, and authorship. I am always most grateful for the support of my loving wife, **Ashley Schenker**, and to my wonderful parents, **Janet and Jeffrey Schenker**.

My family supports me in countless ways by providing emotional support and editorial feedback. Every time I write a book, it's a crazy experience that spills over into my family life, so to them and to everyone else who helped me in this process: Thank you!

Finally, thank you for buying this book. I hope *After the COVID Vaccine* helps you better navigate these uncertain times — and that you find only upside in the wake of this horrific pandemic!

~ Jason Schenker
December 2020

CHAPTER 1

THE NEW, NEW NORMAL

Following the outbreak of COVID-19, countless industries were disrupted. Many began referring to the dynamics impacting e-commerce, remote work, tourism, and other industries as the *new normal.*

With COVID vaccine distribution around the corner, questions about the future are significant. Some wonder if the new normal will stick. Others question if we will wind the clock back and go back to the *old normal.*

So, what comes next?

Whenever I approach this kind of question as a futurist, one of the most important places to look for potential paths forward is the past. Sometimes I'm looking at trends, patterns, and data. For many of the biggest changes in 2020, there wasn't a de novo business, technological, or economic dynamic that emerged on the scene. Instead, 2020 was a year in which COVID-19 became an accelerant for trends that were already in place — trends that I discussed at length in *The Future After COVID.*

In the case of 2021, I've been forced to draw on a certain part of the past to extrapolate what comes next that I would never have expected: my days as a graduate student studying German, including the philosopher Georg Wilhelm Friedrich Hegel. While many people are familiar with the late-eighteenth/early-nineteenth century philosopher's term *Zeitgeist*, which means the spirit of the time, the term *Weltgeist* is less commonly invoked.

Weltgeist translates roughly to the spirit of the world, and in his 1807 work, *Phänomenologie des Geistes* (i.e., *The Phenomenology of the Spirit)*, Hegel argued that this spirit of the world experiences advancements through a dialectical conflict. In other words, the *Zeitgeist* — the spirit of the time — is propelled forward as *Weltgeist* dynamics dictate that one reality gives rise to its antithesis, which in turn yields a conflict that results in a synthesis of the two states.

So, if in true Hegelian fashion, the *normal* of pre-COVID 2019 was our original state, and the *new normal* of 2020 has in some respects become its antithesis, then the technological *Weltgeist* will propel us forward into a synthesis of both the old normal and the new normal in 2021 and beyond. It will be a *new, new normal.*

Hegel's historical approach, which was expounded upon by Karl Marx, is far from perfect. There are many nuances that often get overlooked, like long term trends and developments that accelerated in 2020. But the Hegelian dialectical is a framework that provides a way to think about the future state as a likely blend of what we experienced before COVID and what we experienced during the most severe throes of the pandemic.

In short, the future ahead will very likely be a mixed existence. This is the dynamic I've illustrated in Figure 1-1. For some industries and businesses, this will represent a chance to recover. For others, it might represent a pullback from 2020 activity. But expecting 2021 or 2022 to be just like 2019 would be folly.

At the most extreme upside, the digital transformation forced on the world by 2021 could present tremendous productivity opportunities — and some industries that have experienced growth during the pandemic could remain strong or become even stronger in 2021. On the downside, however, some industries may struggle to lurch forward into recovery or, worse still, stare into an abyss from which they will never return.

Figure 1-1: 2021 as a New, New Normal

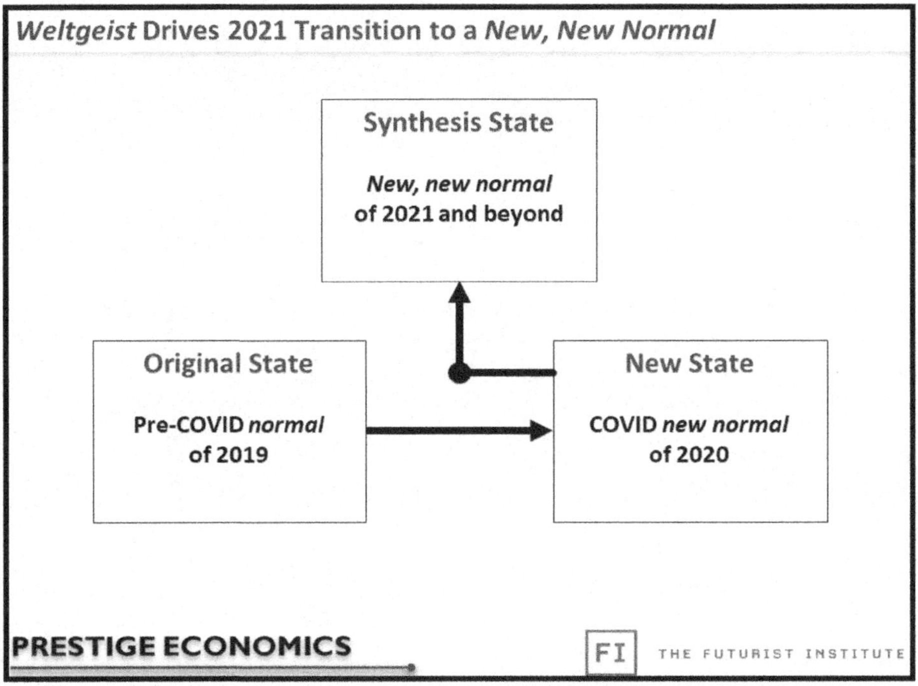

CHAPTER 2

THE SPEED OF NORMALCY

Confirmation bias is the risk that you are looking for data and results that support your expectations. And there are some major confirmation bias risks at play when people make predictions about 2021. After all, while many are hoping for a return to normalcy and to put the COVID-19 plague behind us as quickly as possible, developments could be slower than people want to believe. And as previously noted, many things may never go back to 2019 ever again.

In some ways, this will be a good thing. In other ways, not so much.

Plus, if past is prologue, there is also a risk that timetables of expectations will be pushed back as deadlines fall short. Are we moving forward to a post-COVID world? I expect so. Is it likely to happen as quickly as the latest forecasts reflect? There I am more skeptical. After all, most timetables have so far fallen short. So, I expect plus ça change when it comes to future timetable accuracy.

In a very tangible way, the realities of 2021 will be dictated by the speed of vaccine distribution, which could be slower than people expect. It is unlikely to be quicker. Also, the efficacy of various vaccines "in the wild" could be quite different than in controlled studies. With high efficacy rates for vaccines that exceed 90 percent in controlled studies, there is more asymmetric risk that these could disappoint. After all, for most vaccines on the verge of mass distribution, the upside of 100 percent is only 5 to 15 percentage points away from study efficacy rates.

In short, when it comes to vaccine distribution, adoption, efficacy, and herd immunity, there seem to be asymmetric downside risks. This does not mean that the vaccine won't work or that herd immunity won't be achieved. But it underscores that hopes and expectations are comingled, and hopes are very high.

As someone who has been making forecasts and predictions as a formally trained econometrician since the early 2000s, I would say that there are always risks to being overly confident — especially in timetables that have consistently disappointed. Additionally, I think there is tremendous value in putting percent probabilities around expectations.

So, allow me to spell out my biggest expectations here in black and white. I'll use my anniversary — 2 January — as a date benchmark, as my wife and I often travel around our anniversary. By 2 January 2022, I believe there is a 75 percent probability that we have achieved effective herd immunity through vaccination in the United States and in a number of countries.

Such conditions would be favorable for a more "normal" existence that is accompanied by drastically improved travel, entertainment, and leisure dynamics. I expect to be able to take my wife out to dinner without fear if we do not wear masks. I also expect that we would be able to take a domestic trip for our anniversary without destination restrictions, although masks and other safety protocols may still be elevated. But international dynamics could still be dicey, as U.S. passports might not work everywhere, international travel restrictions might still exist, and cruises might still be restricted.

As these expectations imply, this is something I expect with a 75 percent probability. There is still some chance things could be better or worse by 2 January 2022. It will greatly depend on vaccination effectiveness and whether COVID makes a return for a third wave in the fall or early winter of 2021.

My expectations for a further year out are much more optimistic.

By 2 January 2023, I believe there is a 95 percent probability that we have achieved herd immunity through vaccination in the United States and in a number of countries. I believe strongly that by this time, U.S. passports and international travel options are likely to be more like they were in 2019 than in 2020 or 2021.

For many, this would represent a return to normalcy, even if masks, enhanced sanitization, testing, and vaccination documentation become policies de rigueur for international trips or cruises.

Additionally, while much international travel could be more "like normal," cruise lines could experience long-term financial and safety aftershocks of the COVID-19 pandemic. This means that while cruise lines are likely to be back up and running, some may never recover. And others will have restrictions for a long time to come. I discuss these and other travel dynamics further in Chapter 16.

The Myth of Fauci Claus

One of the most visible U.S. medical authorities during the COVID-19 pandemic has been Dr. Anthony Fauci. Americans of both political parties have revered the expertise and recommendations of Fauci, who is the Director of the National Institute of Allergy and Infectious Diseases. He has served every U.S. presidential administration since Ronald Reagan.

Because so many view him as a man with answers to salve the distressed public, let us consider the potential that Dr. Fauci could, like an areligious Santa Claus, distribute all of the COVID-19 vaccines in one night to every single American man, woman, and child.

Of course, this can't be done, not least because Fauci does not — to my knowledge — have magical reindeer. Plus, and more importantly, it can't be done because the COVID-19 vaccines require two doses spaced apart.

But even if the COVID-19 vaccines could magically be given to every American in one dose, in just one night by Fauci Claus, there would still be side effects.

Plus, there would still be economic aftershocks, like millions of people unemployed, decimated businesses and industries, and systemic technological transformations that would be essentially irreversible at this point.

Given that these aftereffects would still be an issue even if COVID-19 could vanish overnight, we need to consider that these impacts are likely to be even more prolonged in a vaccination distribution and dissemination plan that will not be some sort of 2020 holiday miracle.

Scenarios as a Futurist

Futurists like me think about what the most important levers, drivers, and change agents of the future are. We look at what the big risks and opportunities are, and we examine what trends and underlying fundamentals might not change. Looking at change agents — as well as the big themes and trends that are unlikely to change — helps to foster and lead constructive discussions around what the future may hold.

That was my goal in sharing the myth of Fauci Claus: to uncover the change agents, identify the big themes, the trends, and the immutable realities in order to extrapolate what the potential future might hold as well as what might make it better or worse than projected. In this case, the realities of time, supply chain limitations, side effects, and economic devastation will be the important levers that determine how quickly things become what many of us would hopefully identify as our preferred future: The future after COVID.

Risks of Unbridled Optimism

Before we get to the future after COVID, there are also counterintuitive risks that vaccinations could lead to increased cases of COVID. This isn't because the vaccinations themselves lead to COVID outbreaks, but it could happen because there could be unwarranted unbridled optimism that leads people to throw off the shackles of COVID-19 precaution during the initial time when vaccinations are available for medical providers and at-risk populations, and the later periods when vaccinations are available for a large swath of the population.

In that potentially wide time gap, it is possible that COVID-19 cases could get markedly worse if people become lax in their mask-wearing and social-distancing norms. After all, there is already widespread COVID ennui, and headline-grabbing tales of vaccinations could morph into bad behavior, leading to a worsening of COVID-19 proliferation before mass vaccinations and herd immunity can be achieved.

In many respects, these risks are reminiscent of those soldiers who died in the First World War during the period after which an armistice was agreed to in the early hours of 11 November 1918 and before the time in which the armistice was actually implemented at 11am. During that time 2,738 men died.[1] Those deaths were unnecessary and costly. Yet, they may pale in comparison to the number of deaths that occur in the months ahead because people are tired of waiting for COVID-19 to be over — and they believe that it's time to go back to normal now that the first few people have been vaccinated.

Mutations and Frequent Vaccinations

In addition to the risks that the initial global vaccination will take time, there is also a significant risk that the COVID-19 virus will mutate. As a result, even one round of vaccinations with two staggered shots per person would be insufficient. With mutations, whatever COVID vaccinations are distributed in 2021 could prove insufficient to permanently irradicate COVID.

In other words, such a development could make the COVID vaccine become like the flu vaccine, with annual, modified vaccinations required for efficacy. This does not preclude the potential for 2021 to be a year of vaccination that allows society, business, and the economy to lurch forward to a new normal. But it does mean that society would need to remain vigilant and repeatedly vaccinated well after 2021. Plus, because the negative individual health and public health outcomes for COVID are so much greater than for the flu, this would make COVID-19 vaccinations a top priority as preventative medicine for policymakers, healthcare providers, and insurance companies.

If this dynamic becomes reality, it also means that some elevated social distancing and sanitization protocols for various industries, businesses, and organizations could remain well after 2021.

CHAPTER 3

EVALUATING EXPECTATIONS

In *The Future After COVID*, I included a chapter about weighing impacts. Now, eight months later, there is enough hindsight to evaluate those expectations. And I think it is worthwhile to revisit and revise those expectations.

COVID-19 Impact Size

As I predicted, the human tragedy of the COVID-19 pandemic has been massive. The loss of life, illness, human impact, and economic damage of the COVID-19 pandemic has been as I predicted — catastrophic. Now, looking ahead at the economy, there are reasons to be optimistic. But there could still be more catastrophic second-order impacts ahead. I discuss this in Chapter 4.

Even in *The Future After COVID*, I cautioned that negative impacts for entire industries and segments of the economy could be significantly detrimental for years to come. Of course, I also noted that there was the potential that out of this crisis and tragedy some long-term upside potential for public health, the economy, and society overall may yet be achievable.

Of course, the medical, economic, social, and personal benefits of not having experienced the COVID-19 pandemic would obviously be the best-case scenario overall. But that is not reality.

We are still in the middle of this pandemic. Furthermore, before vaccination distribution reaches scale, we are likely to see further negative public health, social, and economic impacts.

The Future Potential Impacts

With the losses, risks, and costs in mind, we should consider that the potential future impacts of the COVID-19 pandemic could still have some positive long-term impacts. These are all topics I broached in *The Future After COVID*.

Increased remote work is something we have seen in spades since the start of the COVID-19 pandemic. Remote work, as I noted in *The Future After COVID*, offers the opportunity to materially impact the way people work and live. It's been a growing trend for some time, and the COVID-19 pandemic became a threshold moment that pushed remote work sharply forward.

Many people will never go back to an office. And both individuals and companies are likely to be beneficiaries of those changes. The future of work impacts from COVID-19 are discussed further in Chapter 6.

Increased access to online education is also something I highlighted in my book *The Future After COVID*. Indeed, we have seen a significant increase in online education in 2020.

Fortunately, looking forward, increased access to online education can materially impact people's professional and working lives, including the professional and career options they have, as well as the potential future incomes they earn. For many, online education — especially of young children — has been nothing short of horrific. And, at best, it has been tedious.

A friend from high school now runs a kindergarten that she has been forced to run remotely and online. She recently posted on social media how tough it is to tell a child attending kindergarten remotely not run with scissors and have them actually listen to you.

Wow.

Yet, while parents of young children have been forced to do double duty as both remote breadwinners and semi-to-fully present child supervisors, the potential positive economic impacts for online education are significant.

Plus, despite some of the challenges, there is likely to be an increase in the number of home-schooled pupils, which could result in better outcomes for some learners who had not previously considered home schooling or online education.

Of course, the expectations around online education that I noted in *The Future After COVID* were related to postsecondary online education. This is the area where upside potential is significant and postsecondary educational attainment, including degree and certification completion, are likely to increase.

Increased healthcare prioritization is also one of the most important trends I noted in *The Future After COVID*. It is something we have seen at the societal level; it is something we have seen as part of the digital transformation and the leap toward telemedicine and telehealth. Plus, health and medicine are likely to be prioritized as a result of COVID-19 — which could also be the case for a long time to come.

Choices in education, investments, and policies may lead to improved overall public health outcomes. The number of people studying medical and healthcare subjects is likely to increase. Doctors are not where the greatest labor shortages are likely to be; the greatest shortages are in nursing, health aids, and others.

In addition to an increase in health care workers to meet growing demographic needs, the medical supply chain in the United States is likely to be shored up and secured to reduce the likelihood of medical device and PPE shortages in the future. Additionally, we may find individuals become more aware of their own personal health as a result of the COVID-19 pandemic. This could have a long-term impact.

Lower energy consumption and emissions are also something I expected following the COVID-19 pandemic. Indeed, we have seen this in 2020. However, the expectations I noted in *The Future After COVID* were time bound, insomuch as lower energy consumption and emissions seemed likely for a time, as an economic slowdown, remote work, and social distancing norms reduce petroleum fuels consumption. As we look ahead to 2021, energy consumption and emissions are likely to rise again.

The experience of reduced energy consumption and emissions could still have a longer-term impact in how companies view and implement sustainability strategies. But some impacts on tourism and travel could last for some time.

Unfortunately, not all of the impacts going forward are likely to remain positive. One of the greatest negative impacts I noted in *The Future After COVID* is that there is likely to be **greater deficit spending and national debt**. Indeed, since the onset of the COVID-19 pandemic, debt levels have risen globally — and they could rise further still.

Plus, **monetary policy is approaching a risky tripping point**. As I also noted in *The Future After COVID*, there is the growing risk that we are asymptotically approaching a future quantum state of the economy in which the Fed owns everything with nothing at the same time. These risks have also been coming to fruition.

Despite increased fiscal and monetary policy risks, as well as some lasting negative business and economic aftershocks, there is still a significant potential for some positive impacts to be found in the wake of COVID-19 for public health, education, and digital transformation. That was what I expected at the start of the COVID-19 pandemic, and that remains my expectation now. Of course, I still believe that the price we will have paid in the end for the benefits derived from the COVID-19 pandemic will be far too high in terms of lost lives, human suffering, a strained healthcare system, regional and industry economic devastation, a rise in the national debt, and greater central bank obligations.

CHAPTER 4

THE FUTURE OF THE ECONOMY

Looking ahead to the economy of 2021, the most important driver of growth an opportunity, aside from distribution of a COVID vaccine, will be highly accommodative monetary policies. After all, low interest rates are very supportive of growth across wide swaths of the economy. This is especially true for interest-rate-sensitive industries like manufacturing, construction, and housing — although there are some risks to housing due to high levels of unemployment that could persist for years.

Against a backdrop of accommodative monetary policies, there is more uncertainty when it comes to fiscal policy. And while the U.S. presidential election outcome has been settled, important policies related to taxation and government spending need to go through the U.S. legislature. The ability for the Biden-Harris administration to implement policy will be impacted by the outcome of Senate election runoffs in Georgia on 5 January 2021. To control the Senate, Republicans need to win just one of those two runoff elections. Democrats, however, need to win both seats to retain control. Of course, even without Senate control, Biden will be able to set trade, environmental, and other policies.

Manufacturing Strength

One of the areas of greatest strength in the wake of the COVID-19 pandemic has been manufacturing. While services demand has remained significantly depressed, and real GDP is likely to take quarters to return to pre-pandemic levels, there was a relatively short contraction in U.S., Eurozone, and Chinese manufacturing reflected in the purchasing manager indices (PMIs) as seen in Figure 4-1.

This contraction was especially short-lived when compared to the contractions during the Great Recession from 2007 to 2009. This dynamic can be seen in the aggregated sum of the U.S. ISM, Eurozone PMI, and Chinese Caixin PMI in Figure 4-2.

Figure 4-1: Global PMIs[1]

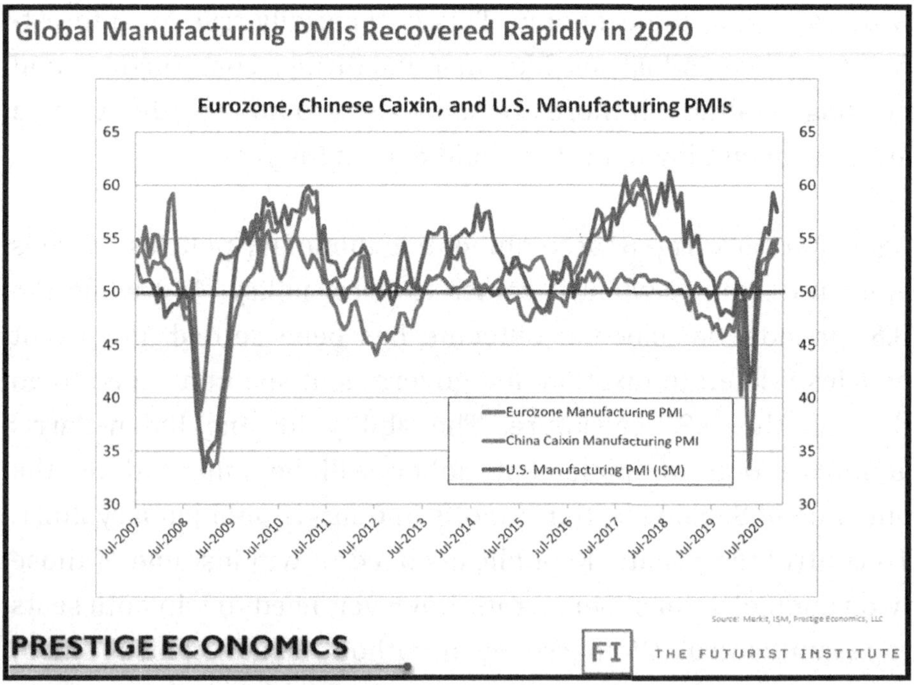

Part of the reason that U.S. and global manufacturing have recovered rapidly and forcefully is because of low interest rates. After all, manufacturing is a tremendously interest rate sensitive industry, and central banks moved rapidly to lower interest rates as the Fed quickly bought up investment grade — and some high yield — corporate debt by expanding its balance sheet.

Looking to 2021 and beyond, interest rates are likely to remain exceptionally low. In fact, forecasts from members of the U.S. Federal Reserve Open Market Committee in December 2020 reflect that the Fed is likely to keep rates near zero into at least 2024. This dynamic is likely to be true globally.

Figure 4-2: Sum of Global PMIs[2]

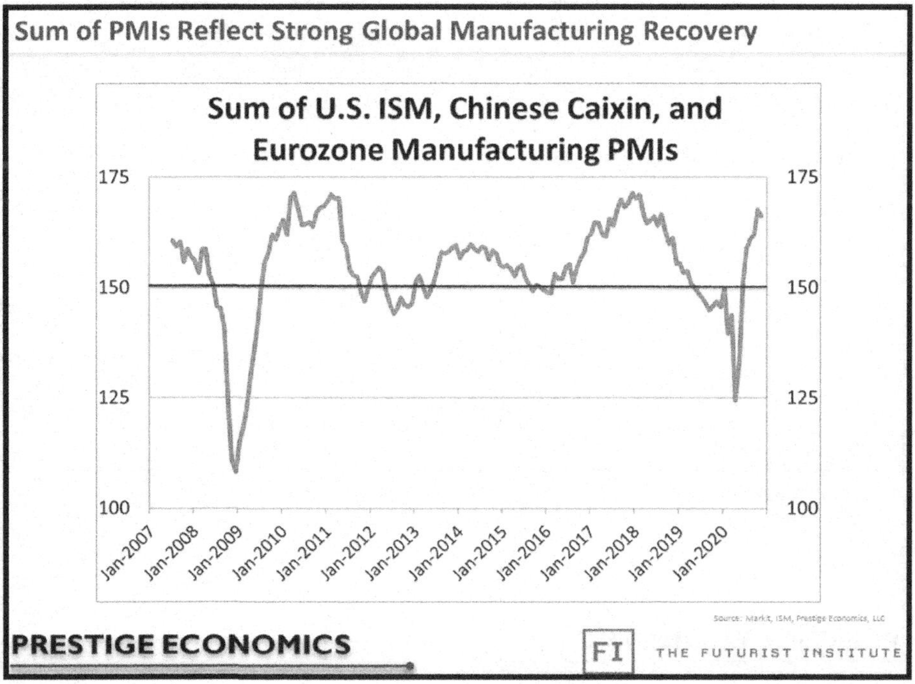

The Fed is also likely to continue buying up mortgages and treasuries in the year ahead. This dynamic, combined with low interest rates globally, is likely to remain generally supportive of manufacturing — and it is likely to be supportive for industrial metals demand and prices.

One area of manufacturing that is likely to remain particularly strong is material handling equipment manufacturing. This critical sector has been supported by a surge in e-commerce as a result of COVID-19. Although e-commerce has been rising on trend for years, e-commerce as a percent of retail sales rose to a record 16.1 percent in the second quarter of 2020. The rise from 11.8 percent in the first quarter of 2020 is shown in Figure 4-3.

Figure 4-3: E-Commerce Percent of Retail[3]

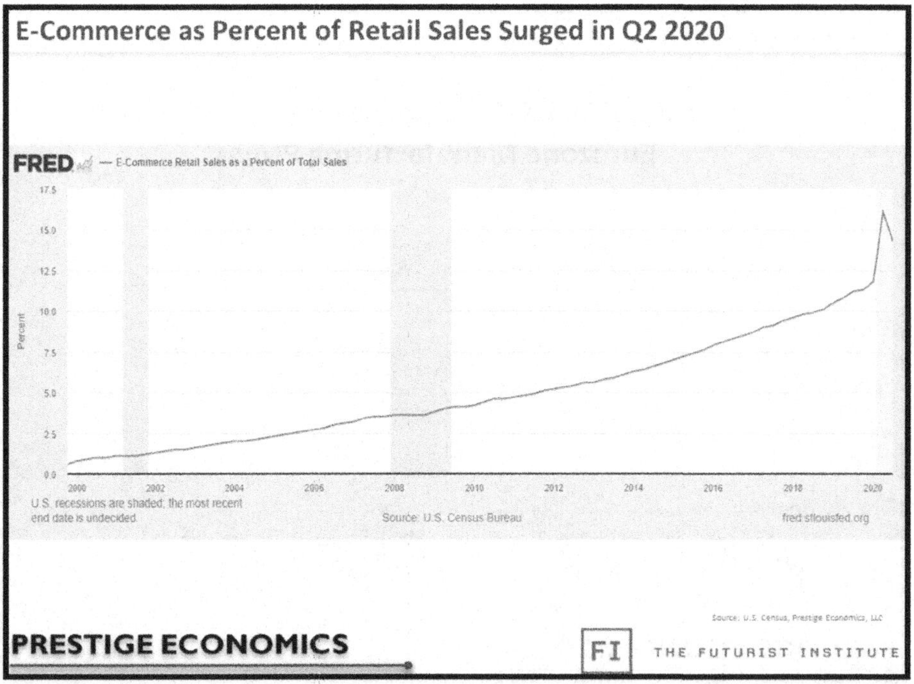

The percentage fell in the third quarter of 2020, when e-commerce as a percent of retail fell to 14.3 percent. Despite this pullback, it did not represent a full drop to the 11.8 percent of e-commerce retail seen in the first quarter. Looking ahead to 2021, it seems likely e-commerce retail percentages could remain above the percentages seen prior to COVID-19.

Beyond supportive e-commerce dynamics, material handling overall stands to be strong in 2021. This is something reflected in Figure 4-4, which shows the percent of material handling industry leaders who expect new orders twelve months in the future to be higher than in the present. This series hit a record level of 100 percent in three months of 2020 through November.

Figure 4-4: Future New Orders for Material Handling[4]

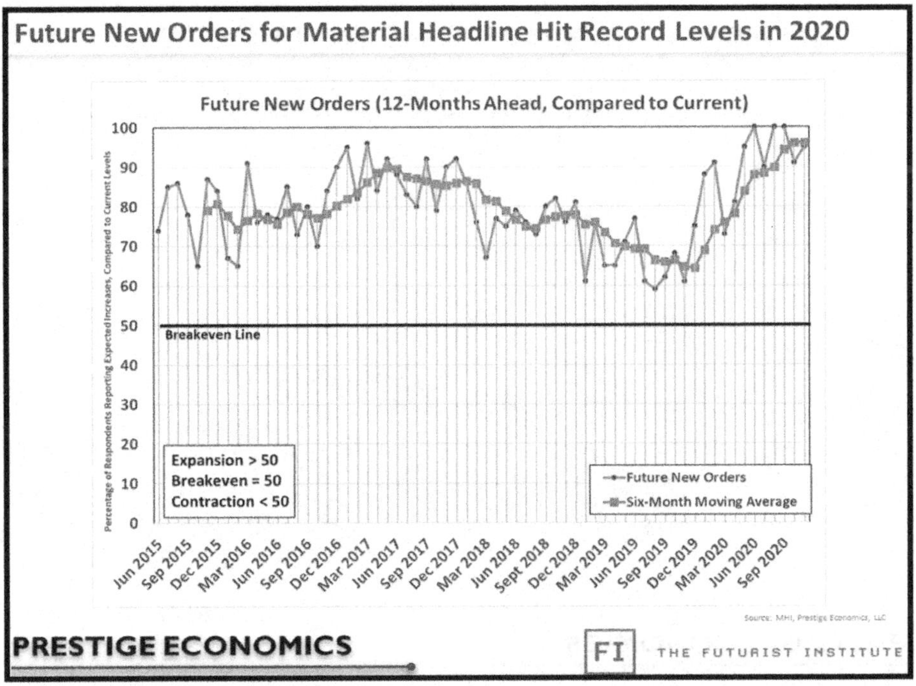

In other words, there were several months in 2020 when 100 percent of material handling industry leaders expected 2021 new orders to be higher than in 2020.

Lurching Toward Recovery

While manufacturing and material handling industries are poised to remain strong in the new, new normal, other industries and sectors could still struggle to recover. Service industries, in particular, are likely to struggle in the year ahead. One area of weakness that is likely to improve significantly in 2021 as we move to the new, new normal is travel and tourism. But this sector has a long way to go, and vaccinations will not be an overnight panacea.

Figure 4-5: TSA Throughput[5]

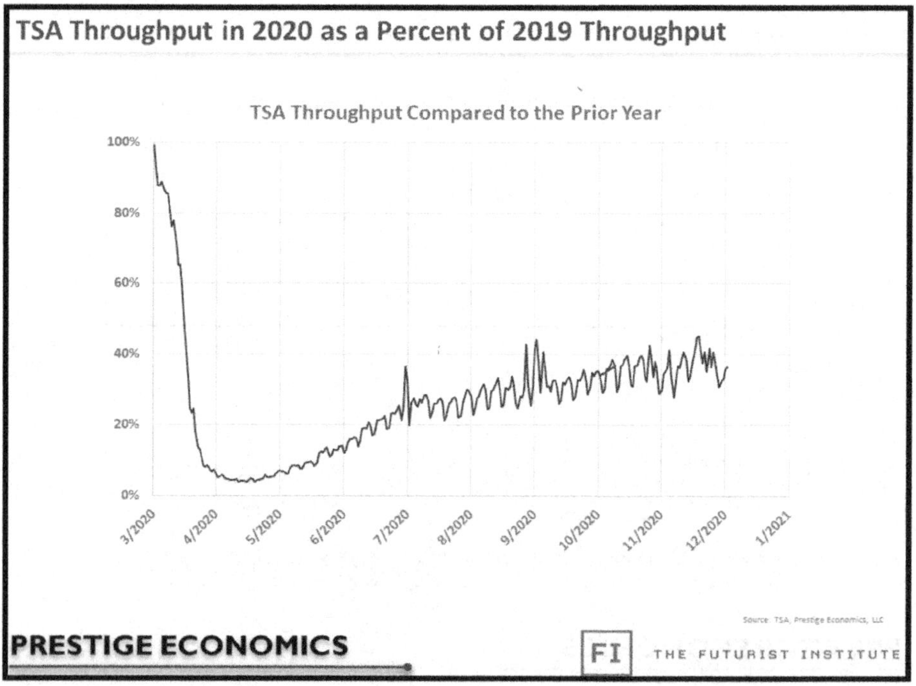

U.S. Transportation Security Administration (TSA) year-on-year throughput has been the best proxy to understand how much travel has taken a hit in 2020. Although down by over 90 percent in April 2020, TSA throughput was still down 67 percent in 2020, compared to November 2019 levels. This dynamic can be seen in Figure 4-5. In 2021, the figure to watch for improvements will be TSA throughput in 2021 compared to 2019 levels.

Tourism and travel could still remain depressed in 2021, as corporate travel and entertainment expenses remain tight, and retail consumers remain both skittish of travel because of the pandemic as well as cost conscious, due to the negative economic impacts of COVID-19.

Along those lines, it is important to recognize that there are a lot of unemployed Americans in 2020, compared to 2019. To see just how many, look at the weekly jobless claims report compiled by the U.S. Department of Labor. While most people focus on initial claims and continuing claims, the best point of reference for COVID-19 dynamics is the table with the persons claiming unemployment insurance benefits in all programs (unadjusted), which is shown in Figure 4-6.

For the week ending 28 November 2020, there were over 20.6 million Americans collecting unemployment insurance across all categories. The prior year, there were less than 1.8 million. That is a massive difference, and while 2021 is likely to be a year of improvements, the job market is likely to be slow to recover, regardless of how quickly vaccines can be distributed. After all, the labor market is notoriously sticky in economic recovery.

Weakness in the labor market could limit some economic upside potential in 2021 — especially for service businesses. But overall GDP growth, manufacturing, and material handling are still poised for solid growth in the year ahead.

The vast array of economic impacts in the new, new normal of 2021 are laid out by industry and sector in the chapters ahead. Recovery for the economy lies ahead, but it will be uneven, rough, and — at times — staccato.

Figure 4-6: U.S. Joblessness in Late November 2020[6]

Over 20.6 Million Collecting Unemployment in Late November 2020

CONTINUED WEEKS CLAIMED FOR UI BENEFITS IN ALL PROGRAMS (UNADJUSTED)

WEEK ENDING	November 28	November 21	Change	Prior Year[1]
Regular State	5,766,130	5,213,781	+552,349	1,751,951
Federal Employees	16,501	13,788	+2,713	9,668
Newly Discharged Veterans	10,202	9,143	+1,059	5,858
Pandemic Unemployment Assistance[3]	9,244,556	8,555,763	+688,793	NA
Pandemic Emergency UC[4]	4,801,408	4,532,876	+268,532	NA
Extended Benefits[5]	693,598	614,517	+79,081	0
State Additional Benefits[6]	3,323	2,363	+960	5,761
STC / Workshare [7]	111,061	101,267	+9,794	9,022
TOTAL[8]	20,646,779	19,043,498	+1,603,281	1,782,260

Source: DOL, Prestige Economics, LLC

PRESTIGE ECONOMICS

FI THE FUTURIST INSTITUTE

CHAPTER 5

BUSINESS PRIORITIES AFTER THE VACCINE

If you run a business, cash is king — especially at times of significant uncertainty. This has been true in 2020, and it is likely to be true in 2021 as well, even as economic conditions improve and COVID-19 concerns fade.

The Importance of Cash Flow

Companies only go out of business for one reason: They run out of money. This is why maintaining positive cash flow and being able to service debt obligations is critical in a downturn. Fortunately, the equation of profitability is always the same, in good times and bad: Revenue minus costs equals profits. As such, there are only two ways to boost profits: sell more or spend less. In good economic times, the focus is often on driving up revenue. And companies are sometimes willing to overlook costs while the money is flowing in.

Travel and entertainment expenses, maintenance expenses, equipment expenses, overtime hours, and other labor expenses can rise significantly as companies rush to capture the upside from increased sales.

As my colleague Nawfal Patel likes to say, "People are too busy making money to ask why or how." However, during a recession or other economic or business slowdown, sales often fall. Sometimes sales stall altogether, as they did for many companies following the COVID-19 pandemic in 2020.

Since increasing sales often becomes tough — but cash flow and profitability are critical — business leaders will often turn their focus on cost cutting as the only other way to support profitability. Of course, this is very tough for some companies. After all, high-growth companies — especially startups — often prioritize growth over profits. Such companies would need to find a way to become profitable after only ever having lost money. That could prove a tall task. But startups are the exception, not the rule.

The good news is that for the great majority of companies that usually show profits, it's a lot easier to keep the money flowing and remain operationally sound in a downturn. Plus, it is a lot easier to strategically reduce costs than it is to shift from losing money in good times to being profitable in bad times. Companies that normally lose money face existential going concern risks during an economic downturn. And even those that are normally profitable can still face risks to their profitability — and their existence.

But their chance of survival is greater. The trick is to keep the money flowing, which is why companies will continue to focus on cash flow and profitability in a number of ways in the year ahead.

This is also why two of the most important plays in the management consultant playbook are to (1) push out payables and pay vendors as slowly as possible, while (2) getting paid on receivables as quickly as possible. Watch for businesses across industries to push these strategies forward in the year ahead as well.

Even though the economy and public health are poised to improve in 2021, risks to business profitability remain high — and some public companies will need to push hard to justify high P/E ratios. So, in addition to focusing on reducing costs and improving cash flow, companies are likely to further embrace technology strategies that demonstrate positive ROI. This includes increased automation in operations, technology as a form of leverage and delegation, and remote work activities.

For some industries that come out of a pandemic-induced hibernation in 2021, like tourism and travel, sales are likely to increase after vaccinations have reached scale and herd immunity becomes a foreseeable reality. But improvements to pre-COVID levels of business activities for these most adversely impacted industries might not be seen until 2022, 2023, or later.

CHAPTER 6

THE FUTURE OF WORK

Remote work surged in 2020. It is something I wrote about with my colleague, Nawfal Patel, in *The Robot and Automation Almanac - 2020*. That book was released in November 2019, and we noted in our essay that the one big thing we expected in 2020 was the potential acceleration in the move to remote work.

We were spot on. But we had no idea how right we would be.

COVID-19 forced many to work remotely in order to preserve the strength of public health systems to prevent loss of life as a second order impact of the COVID-19 pandemic.

In other words, people worked remotely because of government orders to shelter in place or work from home so that hospitals would not become so overwhelmed with COVID-19 patients that they would be unable to treat those with COVID-19 — or with any other deadly maladies or injuries, like from a heart attack, stroke, accident, or other life-threatening situation.

These shutdowns were effective in preserving the ability of healthcare systems to treat people in early 2020. But in the second wave of COVID in the fall and early winter of 2020, hospitals across the United States — and many countries in Europe — became overwhelmed.

As for remote work, it skyrocketed in 2020, with the percent of workers who were working remotely *because of COVID* spiking to over 50 percent of workers with bachelor's degrees, as well as those with master's degrees. You can see this dynamic in Figure 6-1. Interestingly, this figure does not include the percent of workers with either bachelor's degrees or advanced degrees that were already working remotely prior to the COVID pandemic.

Figure 6-1: Telecommuting During COVID[1]

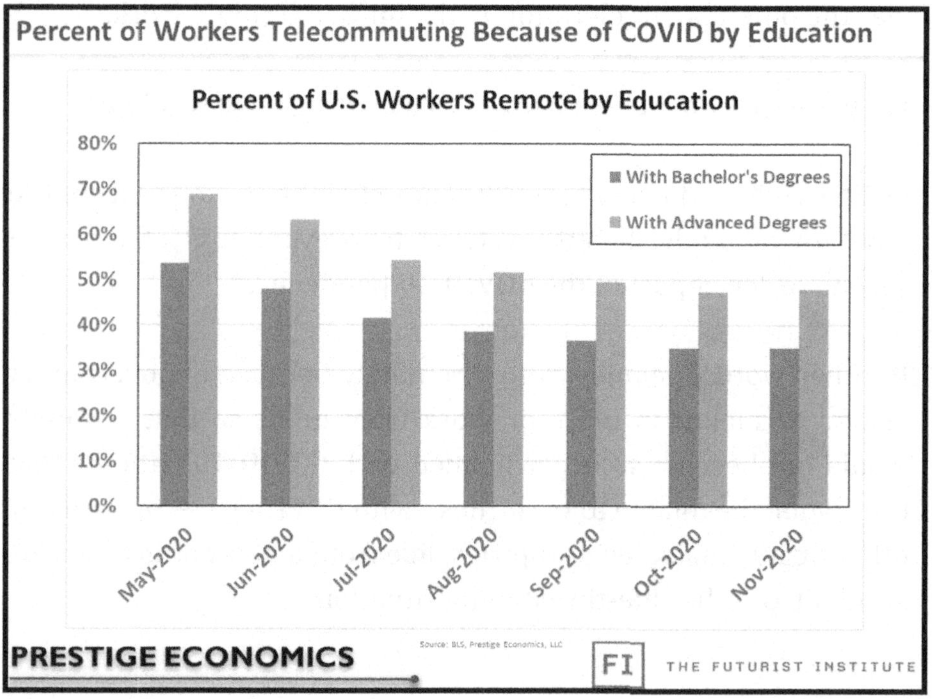

Of course, remote work was not available to everyone. And being able to work remotely has always been highly correlated with education. It's why the peak percentage of U.S. workers with advanced degrees working remotely in May 2020 because of the pandemic was 68.9 percent, while the peak percentage of those with less than a high school diploma who were working remotely in May 2020 because of COVID-19 was only 5.2 percent.[2]

You can see this dynamic in Figure 6-2, which shows those remotely working before the COVID-19 pandemic. For telecommuters, 35 percent had bachelor's degrees and 19 percent had graduate degrees. Those are greater percentages than for non-telecommuters. This remote work potential for knowledge workers surged during COVID-19.

Figure 6-2: Education Levels of Telecommuters[3]

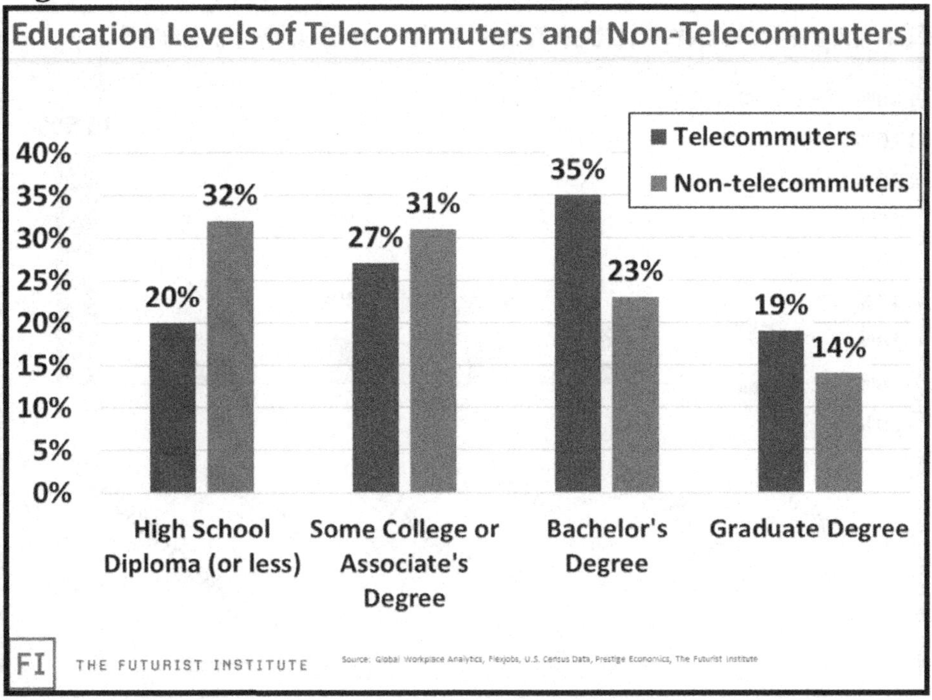

The Trend is Your Friend

There is a saying in finance that "the trend is your friend." So it is, too, with remote work. People have been able to work remotely for well over a decade. This is nothing new.

When I started worked in consulting at McKinsey in the fall of 2007 — over 13 years ago — many consulting firms, including McKinsey, were letting people work from home, already using flex space and co-working spaces. But many corporations have resisted the move to remote working environments. When I founded Prestige Economics in 2009, it was always designed to be a remote work company. We have never had an office. And there have never been any plans to get one.

Figure 6-3: The Rise of Telecommuting[4]

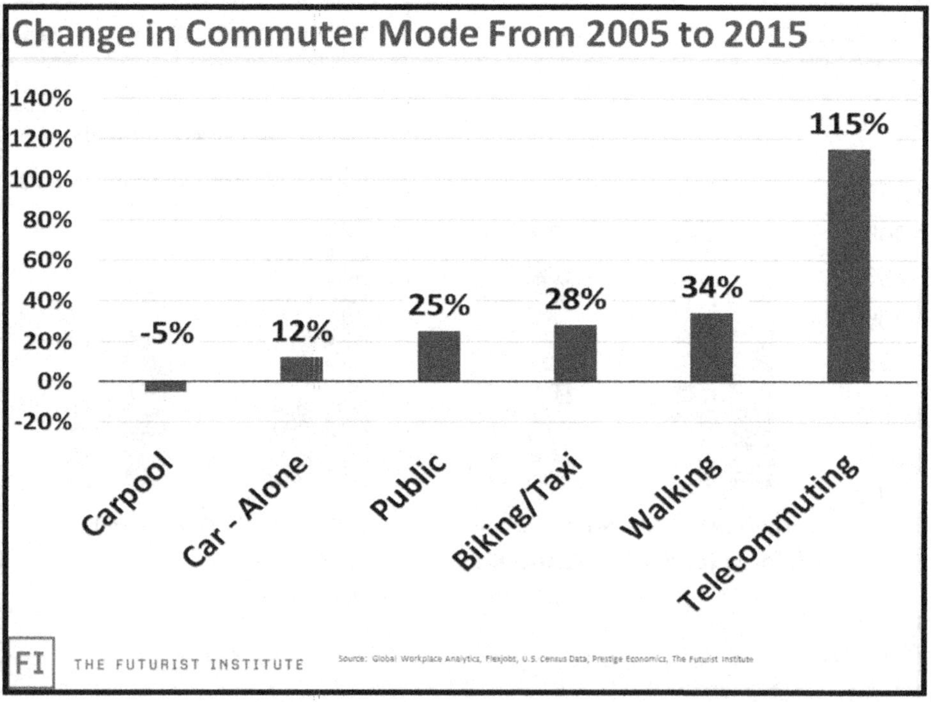

As you can see in Figure 6-3, telecommuting as a means of getting to work increased the most between 2005 and 2015. This growth in the telecommuting sector was already emerging in the mid-2000s, and it is a subject I discussed in my book *Jobs for Robots*.

The value propositions for increased telecommuting — or remote work, in 2020 parlance — are simple. Aside from saving time for workers, it also reduces costs for employers. Employers do not need to spend as much on commercial office space, utilities, paper products, or parking spaces if workers never come into an office. While e-commerce presents upside demand for energy in the decade ahead, the trend of telecommuting presents the opportunity to limit increases in future fuel demand.

Even if the percent of workers telecommuting falls in 2021 compared to 2020 levels, it is wildly unlikely that we will ever see so few remote workers as we did in 2019 ever again. The cat is out of the bag, and there is likely to be a continual long-term trend toward remote work as well as a sustained post-pandemic bump in telecommuting following widespread awareness and experience with remote work.

Looking farther out into the future, telecommuting is extremely likely to continue rising in the decade ahead and beyond. There will be incentives for both workers and employers to make this happen. Furthermore, I expect this will be even more likely in developed economies, where climate change and environmental targets can be more easily met by simply lowering the number of people who drive to work or operate out of the office.

Three Kinds of Workers

One of the big realizations of the COVID-19 pandemic was that there are now just three kinds of jobs — three kinds of workers:

— **Essential Workers**
— **Knowledge Workers**
— **Everybody Else**

First, there are **essential workers**. They have jobs people need to show up for. Essential workers are in healthcare, utilities, manufacturing, agriculture, supply chain, and other critical industries that keep the economy going — and contribute to societal stability.

The second category of workers are **knowledge workers**. They have jobs that can be done remotely. This includes entire industries, like many jobs in technology, finance, and many other professional fields that can operate remotely. Plus, even some office, administrative, and executive staff at essential companies can work remotely.

Finally, there is a third category of worker: **everybody else**. Unfortunately, there are a lot of people in jobs that can't be done remotely — and jobs that are not considered essential. These are a lot of service-based jobs, and these include jobs in restaurants and bars, movie theaters and casinos, hair salons, and nail salons. In total, there are millions of jobs that fall into the nonessential, nonremote bucket. In the United States, my analysis of these job categories yielded the disconcerting potential for 35 million jobs to be lost — or about 22 percent of the workforce.

Of course, some industries did not take a hit in this business cycle. Most notably, those people who could work remotely who had knowledge worker jobs, high levels of education, and high earnings. Plus, unemployment rates and education were inversely correlated in 2020, as you can see in Figure 6-4.

In other words, the more education you have, the less likely you lost your job in 2020. And as companies go into 2021, the war for talent will heat up for those workers with the greatest skills and the most education. In fact, prior to the COVID-19 pandemic, most companies I worked with were concerned about getting the people they needed. Now, as we go into recovery, it will again be top of mind.

Figure 6-4: Unemployment Rates by Education[5]

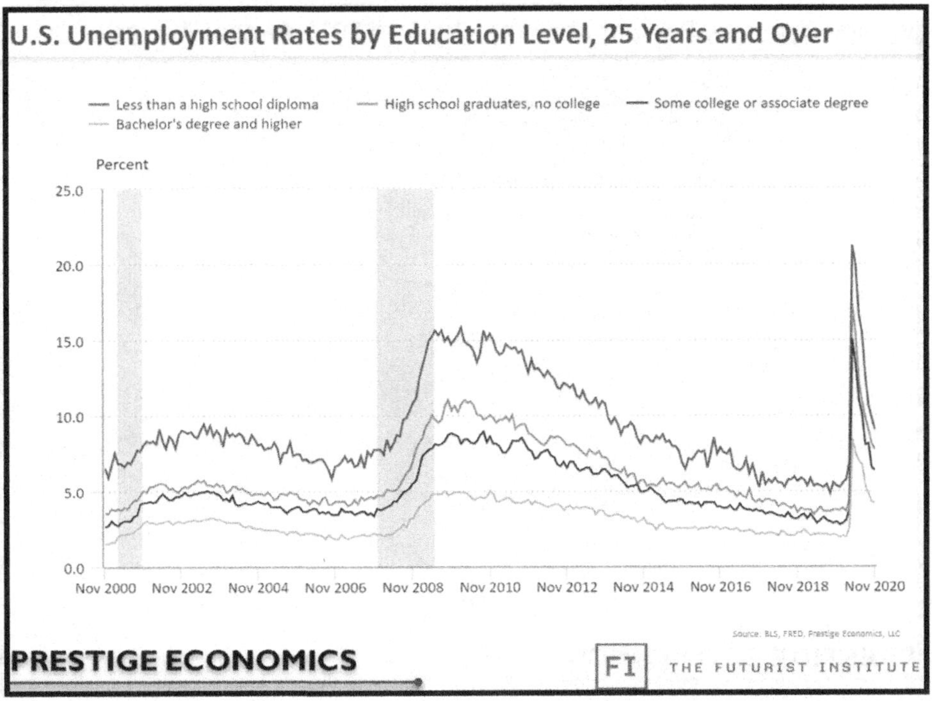

The War for Talent

As we move to a new, new normal in 2021, the labor market is likely to lag behind equity markets and GDP growth in terms of recovery. After all, the labor market is always sticky in terms of declining unemployment rates. But improvements are still likely across industries in 2021.

While nonessential, nonremote jobs are likely to see an uptick in activity in 2021 with vaccine distribution, there are also likely to be increased demands for talent in the fields with the tightest labor markets — and lowest unemployment rates. This includes industries that dominated remote work prior to COVID-19, like professional and managerial roles shown in Figure 6-5.

Figure 6-5: Remote Work by Role[6]

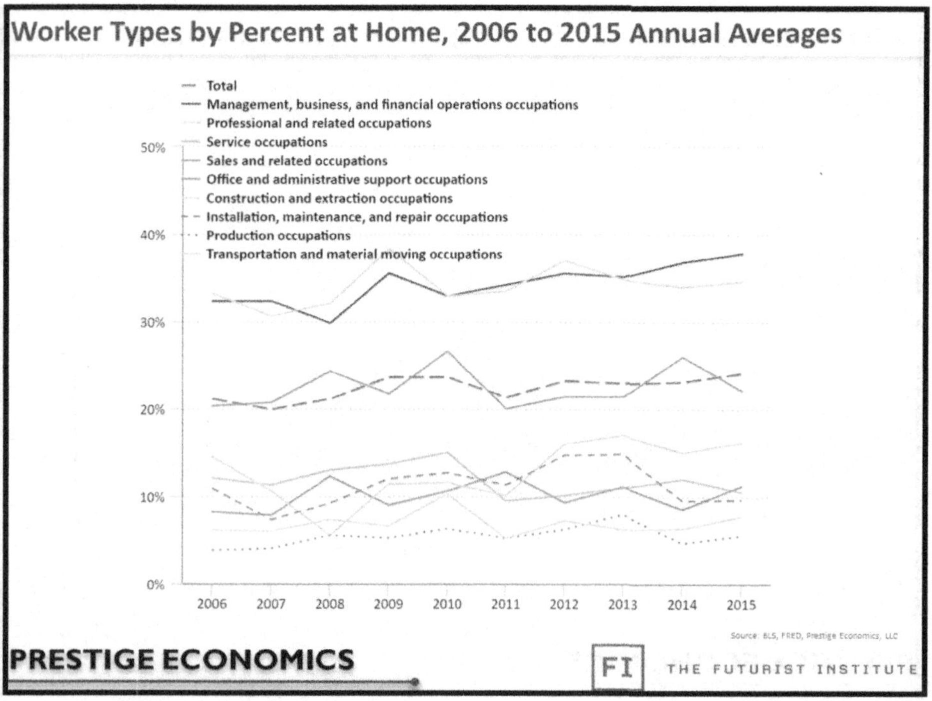

The jobs that were already dominating remote work required the highest levels of education and professional skills, and that will remain true after the COVID-19 pandemic, especially because the majority of workers with advanced degrees and bachelor's degrees have been working remotely because of COVID–19.

While people don't generally like change, the longer they do something, the more it becomes the norm. So it is becoming with remote work, and the longer highly educated people work remotely, the tougher it will be to get them to go back to the old normal of 2019. In fact, going forward, many companies will likely be unable to wind the clock backward.

Some companies will never look back. They will continue to support remote work, which reduces corporate overhead and increases worker satisfaction and flexibility. Those companies will embrace the digital transformation of their workforce and the myriad benefits that accompany it, while managing the downside risks by augmenting remote work with organized, infrequent meetings and offsites, something that will be important for the future of tourism and travel.

These companies will also be advantaged by leveraging remote work for recruiting purposes, effectively turning the global labor market into their local market. Because of the ability to have a workforce tied only to the cloud, I expect these companies will win the war for talent.

Other companies, however, may wish to reverse the trend and go back to a full-time in-office-only work environment.

Of course, now that people are working remotely — and many of them very effectively — companies reversing the trend to end remote working conditions are likely to find it difficult to get their workers back in the office full time.

This is likely to be true for companies with offices in Class A real estate buildings that are the steel and glass dreams of an Ayn Rand character, just as it's likely to be true for organizations the offices of which function as cube farms operating by the low hum of florescent lights.

Companies that insist on only having local talent will find themselves wading into a very shallow talent pool, especially for workers in the highest education tiers — workers who are both in the greatest demand as well as the shortest supply, as reflected by low unemployment rates for workers with master's degrees in Figure 6-6. As with all markets, the price is set by supply and demand. This means workers with the greatest skills and education will have a broader and deeper market to access than ever before.

Organizations with in-office-only work policies may find themselves hamstrung and even existentially threatened in a world where professional and managerial staff have far more options — and leverage in negotiations — than before COVID.

Professions of Growth

Aside from remote professional work, there are several critical categories of workers that are poised for significant growth in the COVID-19 recovery and beyond.

The job categories with the greatest potential for growth and opportunity following the COVID-19 pandemic are health care, supply chain, technology, and project management roles.

Healthcare Jobs

As a result of experiencing the COVID-19 pandemic, it seems likely that people may consider going into healthcare fields. I expect that some young people in college may change their majors after having seen their peers and family members experience negative job market impacts of a COVID-19-induced recession. As other potential professional options dwindle, some students are likely to seek out healthcare to be recession-proof. After all, healthcare is certainly a field well known for that.

Figure 6-6: Unemployment Rate for Master's Degree Holders[7]

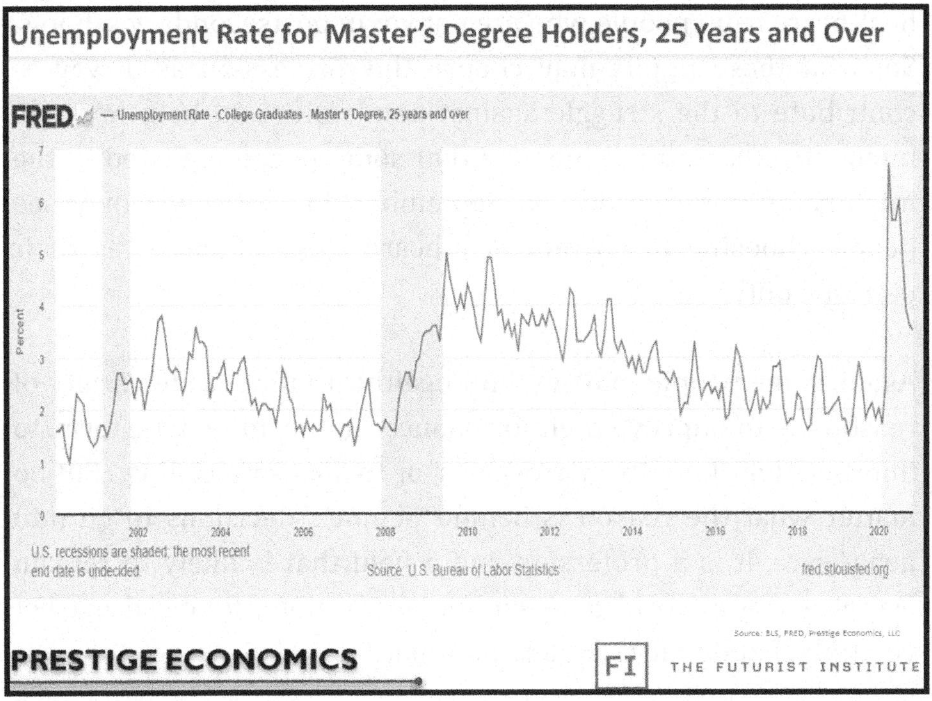

In fact, the recession-proof nature of healthcare is something I have discussed in two of my books back in 2016 and 2017, *Recession-Proof* and *Jobs for Robots*, respectively.

I also expect there will be increased interest in the medical and healthcare professions from midcareer professionals who have lost their jobs or who seek to retool and reskill in order to have more job security in the long run. This makes a ton of sense. After all, healthcare workers are in high demand, and healthcare job categories have long been recognized by the U.S. Bureau of Labor Statistics as critical growth areas in the decade ahead. An aging population, increased longevity, and rising national wealth are all likely to contribute to increased healthcare needs in the future.

Finally, I expect that we will see an increased preference to be in healthcare from people who are not yet in postsecondary schools. Some of these pupils may choose this profession as a way to contribute to the struggle against pandemics and help improve public health. In the same way that some people enlisted in the military after the attack on September 11, 2001, we may see people choosing to go into healthcare as a calling — as their patriotic duty.

Again, some people may get into healthcare for a wide variety of reasons — to improve their job chances in the immediate term, to improve their long-term prospects, or to answer a calling. But no matter what the reason is behind people's decisions to go into healthcare, it is a profession and a field that is likely to remain evergreen in a world of economic uncertainty, financial market volatility, pandemic risks, and automation.

It is also a field that is not funded by discretionary income. After all, if you need medical treatment, you need medical treatment. It's not like the tourism industry, which can see wild volatility around GDP and equity markets. Healthcare is a constant that people will continue to need, regardless of the economy. The sector is also likely to gain further attention in the United States under Joe Biden's presidential administration.

The demographics of an aging U.S. population will also necessitate expanding the ranks of frontline healthcare professionals, including personal care aides, registered nurses, and home health aides. In Figure 6-7, you can see the current expectations for job growth by occupation in the U.S. economy. Healthcare is the clear winner.

Figure 6-7: Greatest Number of New Jobs by Occupation[8]

Most New Jobs: 2018-2028

OCCUPATION	NUMBER OF NEW JOBS (PROJECTED), 2018-28
Personal Care Aides	881,000
Food Preparation/Service Workers	640,100
Registered Nurses	371,500
Home Health Aides	304,800
Cooks, Restaurants	299,000
Software Developers	241,500
Waiters and Waitresses	170,200
General and Operations Managers	165,000
Janitors and Cleaners	159,800
Medical Assitants	154,900

Source: Bureau of Labor Statistics

FI THE FUTURIST INSTITUTE

Additionally, when we think about the potential to automate, it is important to recognize that not all jobs are easy to automate. And healthcare jobs, especially those with high levels of human contact, are likely to be resilient, even in the face of greater levels of automation across the broader economy.

This should not be a small consideration when thinking about jobs in the long run. After all, in the future, many jobs will cease to exist because of automation. This is something driving the U.S. Bureau of Labor Statistics' forecasts for job creation and losses by profession shown in Figure 6-8. As we look at the categories with the greatest growth potential, we again see healthcare is the top winner. It's a need not a want, it's not easy to automate, and demographics favor the healthcare field.

Figure 6-8: Healthcare Job Creation Tops the List[9]

Fastest Growing Jobs: 2018-2028

OCCUPATION	GROWTH RATE, 2018-28
Solar Photovoltaic Installers	63%
Wind Turbine Service Technicians	57%
Home Health Aides	37%
Personal Care Aids	36%
Occupational Therapy Assistants	33%
Information Security Analysts	32%
Physicians Assistants	31%
Statisticians	31%
Nurse Practitioners	28%

Source: Bureau of Labor Statistics

FI THE FUTURIST INSTITUTE

It is the definition of essential. What more do I need to say?

Supply Chain Jobs

Another area of job growth that is likely to continue rising is in supply chain. This is another field that has been growing in prominence in recent years — and it was essential for economic survival during the COVID-19 pandemic. This is the subject of a bigger discussion in Chapter 13.

Supply chain is a field where the need for people is almost as great as the need for automation. After all, the COVID-19 pandemic response revealed that there is a greater need for automation, AI, and robots — especially in the supply chain. Technology in our supply chain was stressed to the limits of its capabilities, as a lifeline of the economy in shutdown. From startups to the biggest e-commerce retail firms, the pressure was on to meet consumer demands — especially with lean inventories.

The COVID-19 pandemic forced us to cross a critical threshold for e-commerce. Now, it is no longer just a convenience; it is a pure necessity. And that will be even truer in the future. Automation will be critical, because people alone could not do everything necessary to meet the economic needs of the supply chain. But looking ahead, I still expect more people will be working in supply chain in the decade to come.

This is despite the fact that the deployment of automated solutions in warehouses and distribution centers will also likely increase exponentially in the years ahead.

Of course, with more automation, the jobs supporting the supply chain will also require greater technical skills. This is likely to be true across industries.

Technology Jobs

For a number of years, I have invoked the aphorism that every company is a tech company — and every job is a tech job. While that may have been hyperbole a decade ago and a bit of a stretch just a few years ago, in the wake of COVID-19, nothing could be further than the truth. The jobs that were non-essential and remote were able to survive — and in many cases thrive — because of technology.

While many may not think of professional and managerial roles as technology jobs, technology is a critical part of those roles. Of course, the need for more applications, networks, data, and other support for companies is also growing. This is where the pure technology jobs are — and they have been booming in the wake of the COVID-19 outbreak. Look no further than the NASDAQ to see this dynamic at play. People working in technology will have many options going forward after COVID. After all, there are only three levers of economics according to Nobel Prize winning economist Robert Solow. Those three elements are labor, capital, and technology. Of those three, technology is likely to become increasingly important — pandemic or not.

Project Management

One final area of significant job growth in years to come will be project management. And this means jobs that help people get projects done and essentially quarterback for specific initiatives.

Some of the permutations of project management also include product management and program management. But no matter what you call it, the goal is the same: stay on top of a team's deliverables and help make sure they get their work done (ideally) on time and under budget.

With more remote workers than ever before, the need for these accountability roles will also increase exponentially in the years ahead — especially for supporting knowledge-worker roles and individual contributors dispersed across time zones who may be operating at the edges of the four corners of the world.

Knowledge Workers and Remote Work

While healthcare, supply chain, technology, and project management roles face some of the greatest potential in the decade ahead and beyond, there will also be rising value propositions for all knowledge-worker roles that can be done remotely.

This is something many people have been working on for some time, but COVID-19 revealed a critical open secret: Being able to work remotely using technology and being a knowledge worker can keep your job alive — even in disruptive times.

To get most roles in healthcare, supply chain, technology, and project management, you need education and skills, which leads us to our next topic: The future of education.

CHAPTER 7

THE FUTURE OF EDUCATION

The future of education is online. That statement was true before COVID-19 as well as during COVID-19. It will also be true long after vaccines have been distributed to every man, woman, and child who wants one.

I've had this expectation in the past, and it's one that I have professionally embraced and continue to hold. In my 2017 book *Jobs for Robots*, I discussed how online education has a chance to significantly better society — and how education is the greatest tool humans have to stay relevant in a world of increased automation.

The Trend is Your Friend

The rising trend in online education has been gathering steam for some time, and now we are seeing that COVID-19 has forced essentially all learners out of classrooms and online. And this is happening at all levels of education — primary, secondary, postsecondary, professional, certificate, and informal.

Of course, there's also been significant growth in online courses on platforms like LinkedIn Learning. And the general trends in massive online open courses (MOOCs) and simultaneous massive online courses (SMOCs) have been growing for a long time. But, of course, this hasn't just been a growth in informal education.

In truth, all of these courses are likely to grow.

Trends tend to build slowly and then accelerate at certain threshold moments. This may be that moment for online education. And the current experience alone may be enough to change the future of higher education, as well as all forms of education, permanently.

I've shared the story in other books about how I completed my third full master's degree entirely online without ever going to a campus. This included defending my master's project remotely, as well as doing teamwork and group projects online.

And I was doing that back from 2014 to 2016.

Of course, the technology is even better now.

The tools people have are better now. Computers are faster now; smartphones are better now. Greater bandwidth is more accessible now. Essentially, everything people need to be the most successful with online learning has all improved drastically in the years since I completed an entire master's degree online.

And as I look to the future of education, I see the potential for three main dynamics to come out of the COVID-19 pandemic experience.

These are the same three dynamics that have affected other industries where technology helps to break a guild system, and education is — especially at the collegiate and post-collegiate levels — a guild system.

That's what the entire model of doing a bachelor's degree, master's degree, and Ph.D. is based on; it's based on the medieval structure of apprentice, journeyman, master.

Those three levels roughly are equivalent to what one would have seen in either knighthood or in medieval formal degrees. The Ph.D., the doctoral degree, is the pinnacle of these, and a dissertation was designed to roughly mirror the *Meisterstück* or *chef d'oeuvre* — the masterpiece — which would make one a *Meister* — a master of their field. Although the words we use in education are different, the structure of higher education is still medieval in nature.

Historically, that guild structure has often been a barrier to entry for many careers and disciplines. But online education has tremendous potential to disrupt traditional studies and the guild of the academy by drastically expanding the reach of course materials and education content. I expect this is what we're going to see at an accelerated pace in the years to come. More people will be educated than ever before online — and in total, because of online education.

Three Trends in Education

The three trends that we've seen in FinTech — cost disintermediation, democratization, and improved user experience — are the same three drivers I expect will impact the future of education.

First is the ability to disintermediate education costs by breaking the education guild system. Second is the ability to democratize access with an online platform and educate more people. And third, there is the chance to improve the learning experience.

Even though most people might think of college or grad school as a time to be in an archetypical suburban or even rural campus environment where lots of the learning process is organically supported by the life experience of being in a certain kind of conducive environment, that's about to change in a big way.

In recent years, smaller colleges have come under financial pressure. Now, with both an economic slowdown and increased move to online education, I expect that we are going to see some small colleges come under enough financial pressures in the near term to shutter their doors.

And while some small liberal arts colleges — or colleges that resist an expansion online — may cease to exist, large universities are likely to see this as a tremendous opportunity to help fulfill and expand their mandate. After all, they will be able to serve many, many more people by increasing access to online education. And they may be able to do so at a price point that more people could afford.

It's important to remember that most people get into education because they want to help people learn. This is true in primary, secondary, postsecondary, and professional education.

The ability to help more people is something that educators would likely see as a greater calling than serving fewer people, provided that the learning actually happens. There is this really critical concern that if you expand the number of learners, that you may actually decrease the quality of the education.

However, the online education experience of the COVID-19 pandemic response is likely to demonstrate that it is possible to provide effective, quality education even while drastically increasing the numbers of students. If this proves true, which I expect it generally will, the biggest and most well-known universities could push to drastically increase the number of students they serve online, which could disintermediate cost.

Education Inflation Outstrips Total Inflation

Education costs have greatly outstripped other costs. You can see this dynamic when examining the education subindex of the U.S. consumer price index (CPI) inflation data when compared to the total CPI.[1]

In Figure 7-1, you'll note that the education costs have almost always outstripped average year-on-year growth rate in total inflation. Needless to say, uninterrupted, this trend is likely to prove unsustainable for the future of education affordability and access. And it's one of the main reasons why online education is likely to explode in terms of adoption.

The ability to serve more people at a low price point and create competition could break the higher education guild and disintermediate cost, providing a massive societal benefit where more people would be able to get more significant education on demand. In *Jobs for Robots*, I wrote about online education as the in-hand classroom, and indeed this is something that has been a potential for some time. Now, we appear likely to see a leap above the previous threshold.

It's also the reason I founded The Futurist Institute back in 2016. And it's why our courses have always been online. We wanted to be able to serve the maximum number of learners possible at a low price point.

Figure 7-1: Education CPI[2]

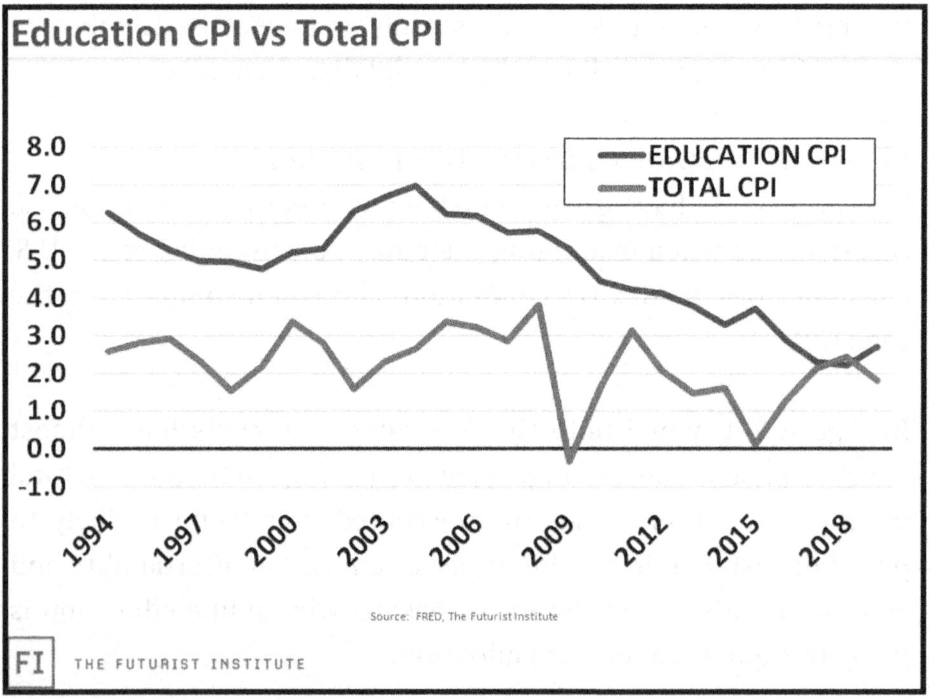

This is how business works, and even though universities are often not-for-profit institutions, they, too, are still businesses of sorts. After all, university budgets must balance — and they also have a mission. But rather than focusing on profits, the mission is to effectively educate as many people as possible.

This is a critical piece, especially when we think about the future of work and we look at the education threshold as being a critical precondition for success as a telecommuter or as a remote worker. In other words, you must be a knowledge worker, and you really need to have a high level of skill or education in order to effectively work online in many cases.

After the COVID-19 pandemic these two dynamics will fold together, where the desire for education will greatly increase because those are the jobs that are pandemic proof. And those will be the jobs that people will want in ever increasing numbers, which means that demand for education will rise since both aspiring knowledge workers and universities will have been forced to cross the Rubicon of online education.

They will likely be able to find ways to continue to serve more people and to more effectively fulfill a broadened scope of their mission to educate people while also increasing the efficacy with which a larger number of people could be educated, at least at a greater way than ever before.

Beyond the disintermediation of cost and this democratization of access, there is also the improved user experience. And this is where we see the technology being a lever.

Paying Elite Tuition Without In-Person Benefits

One of the reasons we may also see a disintermediation of cost in education is because parents who are paying college tuition — or students who are paying their own tuition — to universities to the tune of tens of thousands of dollars per semester, may realize that they can get the same amount of education remotely, just by watching it online.

This realization may push some students and their parents to seek out alternatives in order to get through their courses more quickly, or they may decide that the education value isn't really why they're at the most elite universities.

One of the biggest arguments in favor of elite education is that it introduces you to a network. But that network is a lot more distant when it's remote. And really elite institutions of higher education have been offering online courses for some time.

Harvard has offered courses through its extension school since 1910. And it has been offering online courses and degrees for a number of years at a fraction of the cost of the traditional courses at Harvard University. Of course, there is a distinction between being at Harvard University versus the extension school. But it's probably not that big to most people. After all, it's still Harvard.

And Harvard isn't the only school that offers online courses, remote courses, online certificates, and other programs. Many universities do this. I've personally done certificates online with MIT and Carnegie Mellon. If the content is your priority, then for some learners, the online experience may be most effective.

Of course, online courses force you to miss out on some of the biggest potential of going to an elite institution: networking.

Whether we are talking about online courses, remote courses, or hybrid courses, I expect that the future of education following the COVID-19 pandemic is likely to lead to two critical outcomes.

In addition to expecting an increase in total education attainment at the population level following the COVID-19 pandemic, I also expect an acceleration in the number of explicitly designed online courses. And I believe there could be pushback against some advanced education with very high costs.

After all, people have been more willing to accept the high cost of education in the past because it came with a guaranteed network. But it's very tough to build long-term relationships with influential professors, thought leaders, Nobel laureates, and your academic peers, who may someday be future leaders, if you are only taking a course online that was envisioned, designed, and sold as an in-person value add. The truth is that if the networking is the real value-add (which seems to be the case) and the education is secondary, then the cost of elite university education as well as elite liberal arts college education might suffer some pushback on price, and there could be a preference for networking experience outside of the academy as a substitution effect for the lost networking value.

This isn't a guarantee, of course, but this is just one potential side effect we could see from people watching their children take $20,000 or $30,000 worth of courses online like they're just watching YouTube videos.

The Value Add of Technology

Technological aids are likely to decrease the cost of education and increase the potential for competition. And we also see a significant potential for technology to support democratized access to education and improve the learner experience.

Furthermore, the use of technological aids will only likely improve over time. And if we see universities remain closed for the spring semester of 2021, technological aids that improve the online education user experience could become more critical. At the time this book was published, there was a not small potential for some universities to remain closed through the end of the entire calendar year of 2021, thus canceling any summer or fall semesters for in-person courses.

To support this endeavor, I expect that we are likely to see increased investment in EdTech (education technology) as well as significant investments across tools, training materials, and remote materials that allow for a more cohesive learning process. Once we see students and teachers become accustomed to the online education delivery model using these technologies, we are likely to find people will be more accepting of them.

There's an old German saying, "Was der Bauer nicht kennt, frisst er nicht." It translates roughly as "What the farmer doesn't recognize, he won't feed on." Because guild industries like healthcare and education are based on centuries of tradition, the actual process of getting into their guilds is not necessarily steeped in technology, even in those fields themselves that are immersed in technology and embrace technology.

I expect that we are likely to potentially see some changes stemming from technological disruption to these guilds in the decade ahead — and beyond. The unique situation of COVID-19 may even hasten the process, as it has revealed a shortage of healthcare workers, while also bringing to light the tremendous potential for online education and remote work.

There are also likely to be important regional opportunities, In the state of Texas, where I live there is an initiative known as 60 by 30; the goal is to have 60 percent of Texans age 25 to 34 complete some level of postsecondary education, degree, or certificate by the year 2030. It is an ambitious goal, but it is critical because the jobs that are remote and less automatable are essential for the next economic phase of growth in Texas. And it is exactly this cohort of jobs that requires education and technological skills.

Trade skills, formal education, informal education, and perpetual learning are all likely to gain a boost from online education. And the rise in online education due to the U.S. COVID-19 response makes Texas much more likely to achieve the 60 by 30 goal.

This is not to say that the pandemic of COVID-19 is positive. It is most certainly not. It is horrific. It is a catastrophe.

Yet, if we were to consider the impacts for the economy and society at the population level, we might say (in an attempt to find kernels of hope in a desperate situation) that now maybe in the long run our populace will be more educated — and we will have a much better prepared workforce for the decades ahead.

From an economic standpoint, this is invaluable. From a population stability standpoint, this is invaluable. And from a public health perspective, where we see there is a massive deficit of people in healthcare fields — and a massive need going forward to fill healthcare jobs — we also expect that online education will help to facilitate filling those roles in the healthcare fields going forward. If that is a result, then public health outcomes in the long run could be greatly improved.

This is a topic for discussion in the chapters about "The Future of Healthcare" as well as "The Future of Work." But it's really important as we think about the future of education to bridge that gap between the needs of the economy, the needs of the workforce, the needs of the people and public health needs, and the educational capabilities to have the workforce cross a divide that up until now seemed honestly almost insurmountable and quite daunting.

I am much more optimistic now that we will achieve improved levels of public health over time because we may be able to create not only a more educated workforce, but we are also likely to see the education enrollment in health sciences and life sciences increase in a way that may also result in a long-term positive net benefit for public health and the economy.

Again, this is not to say that COVID-19 is in any way a positive. But if we were to look at long-term ramifications, there is reason for hope that out of this pandemic tragedy and economic crisis, we might be able to derive something of value and something positive in the long run.

The Impact on Home Schooling

One dynamic that has yet to be determined is what the experience of involuntary at-home, in-hand learning will be on homeschooling. In Figure 7-2, you can see the trend in home schooling since 1999. Both the absolute level and the percent of U.S. pupils aged 5 to 17 who are home-schooled appear to have peaked in 2012 before slowing in 2016.[3]

For many people with young children, homeschooling has been a major challenge. Nevertheless, I know many people who intend to continue to homeschool their children. As I noted in *The Future After COVID*, it seems quite likely that the COVID-induced widespread mandate for home schooling might result in an increase in home schooling in the United States.

Figure 7-2: Home-Schooled Pupils[4]

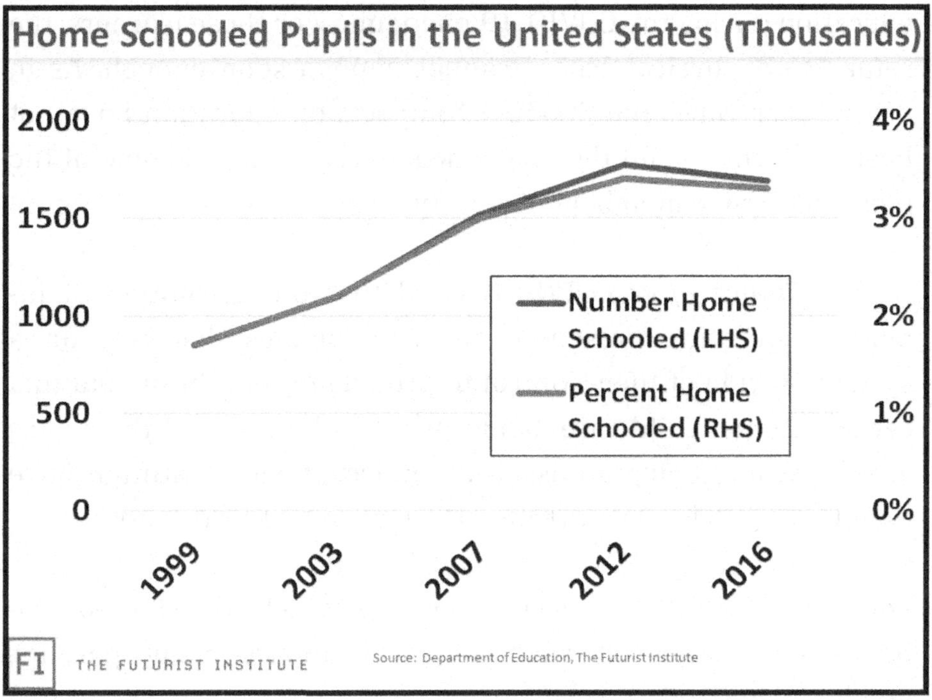

After all, the new experience for many may actually result in better outcomes for some learners. Plus, the percentage of home schoolers is so low that is almost impossible to imagine that the number would not go up as a result of an experience that may benefit some learners.

Meanwhile, pupils who were home-schooled before COVID-19 are unlikely to turn away from home schooling as a result of COVID-19. In essence, their educations and learning lives are likely to be much less disrupted than pupils in public, private, charter, religious, and other physical schools. I mean, why would they want to stop doing the only thing that works?

Of course, some learners without adequate internet access or technological hardware have suffered greatly from remote education during the COVID-19 pandemic. For those learners, the return to a physical school cannot happen soon enough. As in many other ways, the COVID-19 impacts on education have not been uniform — and they have negatively impacted some of the most vulnerable members of society.

In *The Future After COVID*, I noted that the disruption of in-person education presented risks to families that rely on a system outside of their control to provide for education and to a certain degree child care. Some people who planned their lives around school being an assured, undisruptable institution have found the COVID-19 pandemic to be a catastrophic disruption.

Long after COVID-19 vaccinations are widely distributed, the impacts of disrupted education will impact many parents, families, and pupils.

Regardless of the disparate experiences people have had during the COVID-19 pandemic, more remote learning and at-home, in-hand education is likely to be seen across all learning levels, from home-schooled primary school pupils to undergraduate college students and doctoral candidates. On the upside, increased access to all levels of education is likely to result in improved economic outcomes at the population level.

Historically speaking, education has been the great divider for jobs. In Figure 7-3, you can see statistics on unemployment and earnings from 2018 from the U.S. Bureau of Labor Statistics. Education is positively correlated with income, and education is also inversely correlated with unemployment. So, the more you learn, the more you earn — and the less you are unemployed.

Figure 7-3: Education, Earnings, and Unemployment[5]

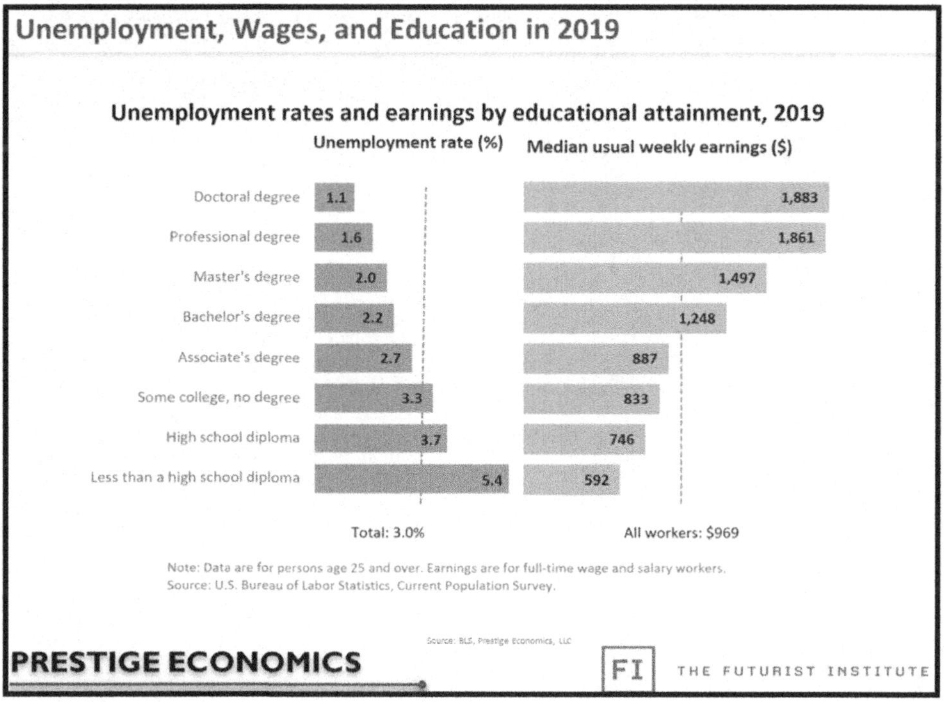

This is likely to be just as true after COVID as it was before or during the pandemic. And as we saw in Chapter 6, the jobs of the future will require education and skills. It is my expectation as well as my hope, that the increased access to online education will allow for more people to advance in their careers and improve their lives for a long time after the COVID pandemic has ended.

CHAPTER 8

THE FUTURE OF ENERGY

When people debate the future of energy, electric vehicles is a hot topic — and it's bearish for crude oil demand. But it ignores one much more critical factor: Global wealth is rising, and petroleum-fueled vehicles are going to become more important with the expansion of emerging market middle classes in developing economies where energy poverty is still an issue.

This means that the locus of critical oil demand is also going to shift from the OECD to the Far East and other emerging markets.

In the same way that U.S. shale oil production presents a dynamic shift in terms of the geographic source of additional marginal barrels of crude oil, the rise of the middle class in emerging markets presents a similar risk for the future of marginal additional oil demand. So, while the 2020s are likely to be a decade in which U.S. crude oil production and refineries play an increasing role in the global supply chain of crude oil, it is also likely to be a decade in which Chinese, Indian, and other Asian demand rises drastically.

Telecommuting and Crude Oil

In 2010, I attended a meeting of the International Energy Forum (IEF) in Cancun, Mexico. The IEF is a venue where the world's biggest oil producers and consumers have a discourse about the future of energy. It is also an organization that shares energy data between countries. Think of it as OPEC and other oil producers — plus the biggest consumers in the world of crude oil.

At that event I gave a television interview for CNBC on the beach in a linen jacket. It was quite ridiculous in hindsight. Anyway, the interview was with Carl Quintanilla, who was filming a segment for his special series titled *Beyond the Barrel*. Again, this was back in 2010. He asked me about the prospects for renewables and electric vehicles and how that would impact crude oil prices.

My response to him then was the same as it would be today. On the power side, there is tremendous potential for renewables like solar and wind to supplant coal power generation. However, on the transportation side of the equation, there really isn't a more efficient form of transportable energy than liquid hydrocarbons. And the limitations of battery technology are significant, from both a technological and elemental perspective. In other words, there aren't enough elements to replace all of the conventional vehicles in the world.

He asked me what oil producers like OPEC should be worried about, and I noted that telecommuting presented the greatest downside risk to oil prices in the future. After all, I had already been working remotely for three years at that point — and I didn't see a reason why others couldn't work remotely as well.

From an energy standpoint, if you don't have to leave your home to go to work, you don't need to drive and you don't need to have two spaces — your home and your office — cooled or heated to account for climate control.

As I discussed in Chapter 6, remote work is not a new trend — and its one that has gained tremendous momentum in the past decade and a half. Apparently, I'm not the only one who has been working remotely! And that was most certainly true in 2020.

COVID-19 "shelter in place" rules were a de facto mandate of petroleum fuels demand destruction — especially gasoline and jet fuel. They also allowed us to see the impacts telecommuting could have on reducing oil demand. Plus, we also see the potential ESG impacts of such policies, something I will discuss further in Chapter 17.

As we push toward a new, new normal in 2021, crude oil and petroleum fuels demand are poised to rise. This is also likely to support prices. And those long-term trends of rising global population and per capita incomes in emerging markets are likely to support energy consumption in the long run. But remote work is likely to remain critical for energy prices in the medium term.

Remote work seems unlikely to be a dynamic that can fully counter the future rise in energy and fuel demand over time due to significant expansions in the global population and a likely massive rise in emerging market global wealth. But telecommuting may provide some future demand abatement that also limits oil price upside and satisfies ESG mandates.

CHAPTER 9

THE FUTURE OF MONETARY POLICY

Once upon a time, long before COVID, there were central banks that had balance sheets that merely facilitated their day-to-day operations. Sometimes, they raised interest rates to dampen business and fixed investment. Other times, they lowered interest rates to encourage business and fixed investment.

Central banks used deposit rates, interest rates on excess reserves, and other interest rates (like the U.S. federal funds rates) to impact economic growth. But they didn't go into financial markets and buy assets as a means to influence interest rates or keep the economy from coming apart at the seams.

Then, one day, the Great Recession happened. A housing bubble in the United States spilled over into the financial system, causing a crisis. After dropping rates to zero, the only way to revive housing appeared to be by having the central bank buy up mortgages. At the same time, buying government debt also appeared to be a solid way to dampen interest rates to encourage companies facing mortal danger to take risks and make investments. This, too, happened many years before COVID-19.

Well, this surprising plan worked. Boy, did it work. The Fed, European Central Bank, Bank of Japan, and others expanded their balance sheets by essentially pulling money out of the sky to buy assets with maturity dates. The logic was that this short-term buying with money that previously did not exist wouldn't be a problem, because central banks could allow these fixed income products of mortgages, treasuries, and other debt assets to mature and roll off the balance sheet. When that happened, they would turn back into cash — cash that never existed — and vanish altogether. The U.S. central bank, the Federal Reserve, expanded its balance sheet from under $900 billion before the Great Recession to over $4.5 trillion by December 2014, as you can see in Figure 9-1.

Figure 9-1: Fed Balance Sheet[1]

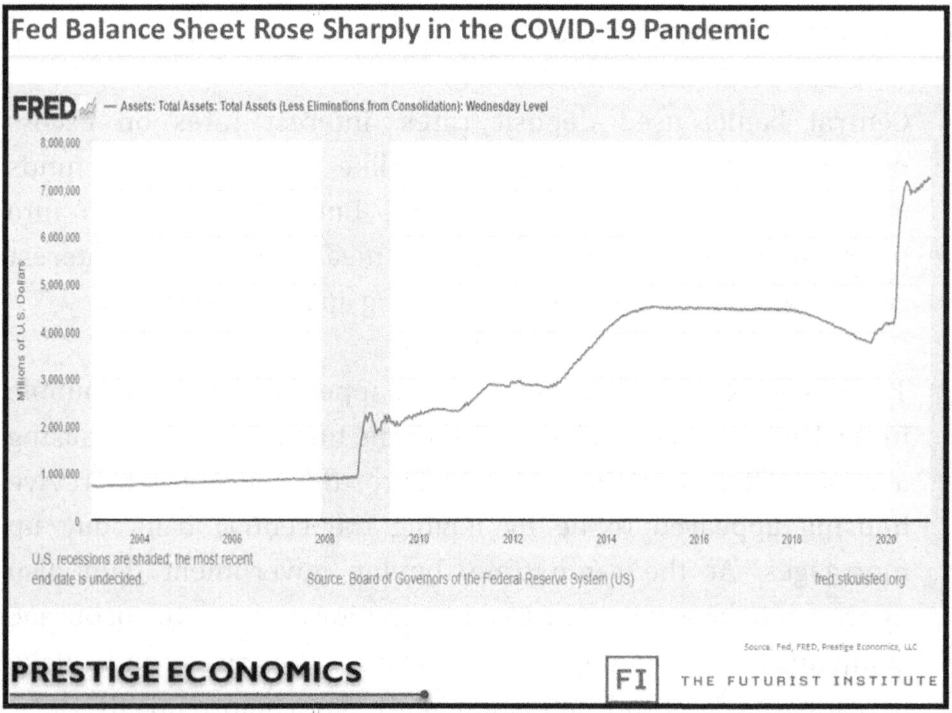

The ECB also expanded its balance sheet by over 1.5 trillion euros through mid-2012. Then, the ECB — the progeny of the fiscally conservative BuBa — tried valiantly to shrink its balance sheet between mid-2012 and late 2014 by about 1 trillion euros. It failed miserably. Of course, the balance sheet shrank, but the Eurozone was almost pushed into a recession. And the ECB had to reverse course and begin rapidly expanding its balance sheet again.

The Fed would also eventually try to shrink its balance sheet (in much smaller percentage terms) during 2018 and 2019. That plan also failed. In fact, the Fed's balance sheet reduction helped cause a business and fixed investment recession in the United States in 2019. This is why before the COVID-19 pandemic, the Fed's balance sheet was already re-expanding.

Given the economic aftershocks of these two failed attempts to reduce an expanded central bank balance sheet, it appears as if these balance sheets can only grow. They cannot shrink. So, when the assets mature, the funds need to be reinvested into financial markets — or economic growth will slow sharply.

Then, in 2020, in response to the COVID-19 pandemic and recession, central banks again massively expanded their balance sheets. Central banks began buying a much broader range of assets, and this was completely foreseeable. In fact, Janet Yellen even noted at the annual Kansas City Fed event in Jackson Hole, Wyoming, in 2016 (when she was the Fed Chair) that "asset purchases will remain important components of the Fed's policy toolkit."

She further added that "future policymakers may wish to explore the possibility of purchasing a broader range of assets."[2]

In the wake of the COVID-19 pandemic outbreak, that is exactly what happened and the Fed's balance sheet in mid-December 2020 was around $3.1 trillion greater than it was on 1 January 2020. Along with a massive ballooning of the Fed's balance sheet, U.S. federal funds rates are close to zero as shown in Figure 9-2. But this may not be the bottom for U.S. interest rates. Looking ahead to a time after the COVID vaccine, rising government expenditures and income shortfalls could push the Fed to enact negative interest rates.

Figure 9-2: U.S. Effective Federal Funds Rate[3]

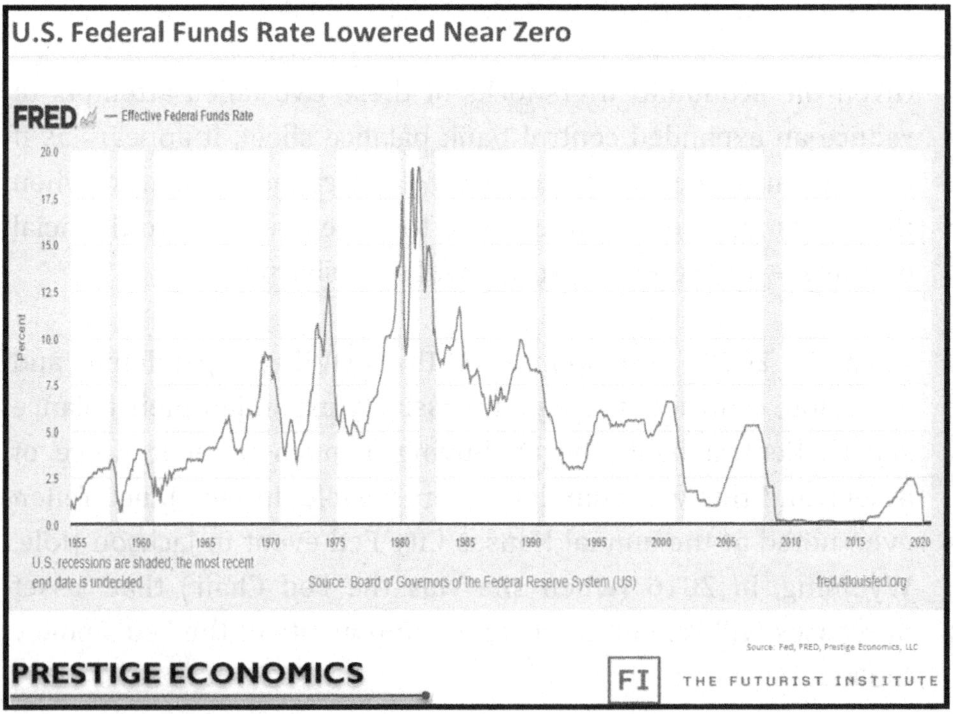

It sounds shocking, but there have been negative rates in the Eurozone since 2014 — against the backdrop of the ECB's failed attempt to reduce its balance sheet.

Policymakers refer to the use of expanding balance sheets as quantitative easing. But whenever I speak about this topic to clients or in a public speech, I feel like I'm discussing some crazy, fantastical, magical fairy tale. Sometimes I even call it magical monetary theory. In short, that's not a good sign for sustainability. The question now is, how long can these dynamics reign?

Central Bank Balance Sheets and Cryptocurrencies

The expansion of central bank balance sheets is fundamentally supportive of the economic and financial arguments in favor of Bitcoin and digital currencies that are not backed by central banks. In fact, the first Bitcoin transaction, which is often referred to as the Genesis Block, included the following message:

The Times 03/Jan/2009 Chancellor on brink of second bailout for banks[4]

The Bank of England engaged in multiple bailouts and a 300 percent expansion in the size of its balance sheet between 2007 and 2012. The impact of these expansions and persistently high levels of central bank balance sheets of the Fed, the BoJ, the ECB and the Bank of England is unclear. What has been clear for some time, however, is that central banks have cracked open the QE cookie jar — and that they were likely to spend more money they don't have, effectively creating the ability to purchase assets out of thin air long before COVID-19 became a risk. And this will remain true after COVID vaccine distribution.

There is no problem with central bank balance sheet expansion from an accounting standpoint. After all, if the assets eventually mature and expire in value, they could fall off the balance sheet. But that is not going to be the case with the balance sheet of the Bank of Japan, which includes significant equity assets. And that also seems increasingly unlikely for the ECB, the Fed, and other central banks.

As we look ahead to the future of finance and the future of quantitative easing, one of the things I often speak about is a risk that with each cyclical downturn in the future, central banks may continue to expand their balance sheets. As long as they are all doing this, however, it may not drastically impact foreign exchange rates. After all, if everyone plays the game, it's tougher for there to be an outright winner — or loser. Plus, they will all want to play the quantitative easing game again if they can, just as they did after the Great Recession and in the wake of the COVID-19 recession.

The Future Quantum State of the U.S. Economy

The dynamics I have just described are likely to persist well after vaccines have been distributed. In fact, they could eventually result in something I refer to as the future quantum state of the economy.

As the U.S. Federal Reserve buys more assets during each downturn with money that it pulled out of nowhere, the central bank's balance sheet will grow and grow.

With each cycle, the central bank will become increasingly important as the buyer of last resort so that almost any company or municipality in the entire U.S. economy becomes too big to fail. This could also become extremely challenging, as the U.S. national debt rises due to unfunded entitlements and interest.

One potential worst-case scenario is where the central bank — after decades of cycles — eventually owns almost everything in the economy. And it will have paid for the debt, MBS, Treasuries, equities, and maybe even physical assets on its balance sheet with money it created — with money it didn't have in the first place.

This is effectively how we could have a quantum state of the economy, where the central bank owns everything with nothing. Then we would have a very big problem.

Some analysts have often wondered out loud if there must be a limit to the size of the balance sheet of the Fed or the ECB or the Bank of Japan. But maybe not.

Then there's the question of negative rates. I mean, who's buying bonds that cost them money?

Honestly, that is a great question.

Time will tell what comes next. But in the world of economics and finance, magical and fantastical off-balance sheet entities are quite inconsistent with durability and sustainability. Still, as traders like to say, markets can remain irrational longer than you can remain solvent.

In terms of options, one thing seems clear. Central banks are coming to the end of the line. Even if rates go negative, they seem unlikely to go significantly negative. After all, rates are already on top of zero. This means that the biggest stimulative policies for asset valuations from both quantitative easing and lower interest rates may already be behind us.

But negative rates could still lie ahead after the COVID-19 pandemic, after the vaccine, or at some point in the next business cycle.

Fed Chair Powell was once criticized for using a navigation analogy in the face of uncertain policymaking. But it was apt enough. Of course, the level of risk to markets, policy, and the economy has significantly ratcheted up since then as a result of the COVID-19 pandemic and recession. And as any good navigator will tell you, when venturing into uncharted territory, it's best to heed the warning of cartographers of yore: Here be dragons.

That is true no matter how quickly — or slowly — COVID vaccinations are made available at scale to the general public.

CHAPTER 10

THE FUTURE OF FISCAL POLICY

In the wake of the COVID-19 pandemic, economic shutdown, and recession, the global economy has been kept afloat by a mix of monetary policy and fiscal policy stimulus. While monetary policy is set by central banks that have been enacting policies to depress interest rates, fiscal policy has been driven by governments issuing record levels of debt.

Even with vaccine deployment, more government debt could be required to keep economies going in 2021 as the world lurches — rather than rushes — forward to the new, new normal in the year ahead.

The National Debt

The U.S. federal debt has risen sharply in recent decades. The impact of COVID-19 made the situation drastically worse. The official federal debt figure from the U.S. Department of the Treasury at the end of Q1 2020 was $23.2 trillion. By the end of the second quarter of 2020, however, the national debt had risen to almost $26.5 trillion.

In other words, in the second quarter of 2020, the U.S. national debt rose by almost $1.1 trillion *per month*. I noted my expectations for this kind of impact in *The Future After COVID*. But the debt hasn't stopped rising, and in the third quarter of 2020, the national debt rose even further to $26.9 trillion.
That is a lot of debt!

As you can see in Figure 10-1, the pace at which the U.S. national debt is rising has accelerated in recent years. It took 205 years for the U.S. national debt to exceed $1 trillion, which happened in October 1981. But it then took less than five years for the national debt to double to $2 trillion in April 1986.[1]

Figure 10-1: U.S. Debt[2]

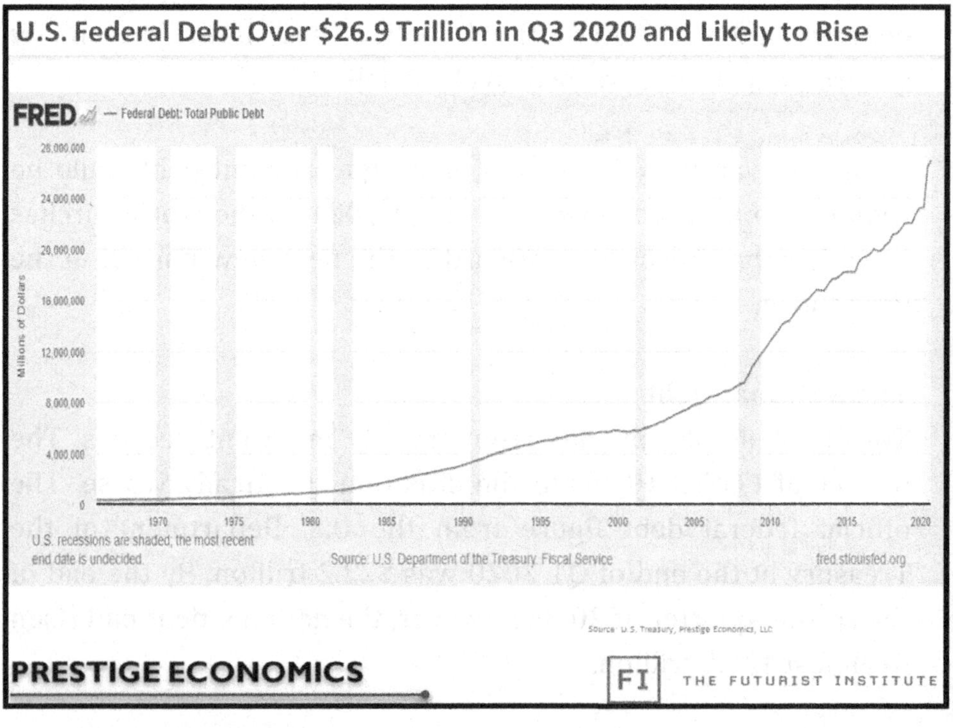

Put another way, while it took 205 years for the national debt to exceed $1 trillion, that's roughly how much it grew during each month of the second quarter in 2020. And there is likely to be a lot more debt to come to revive the U.S. economy from the worst labor conditions in its history.

Although not as pronounced as the trend in federal government debt, the U.S. debt-to-GDP ratio has risen sharply since the onset of the Great Recession as well (Figure 10-2). One major negative impact of a high national debt is the drag on potential future U.S. economic growth as measured by gross domestic product — or GDP. Plus, debt exposures can be exacerbated by compounding interest on already outstanding government debt.

Figure 10-2: U.S. Debt to GDP[3]

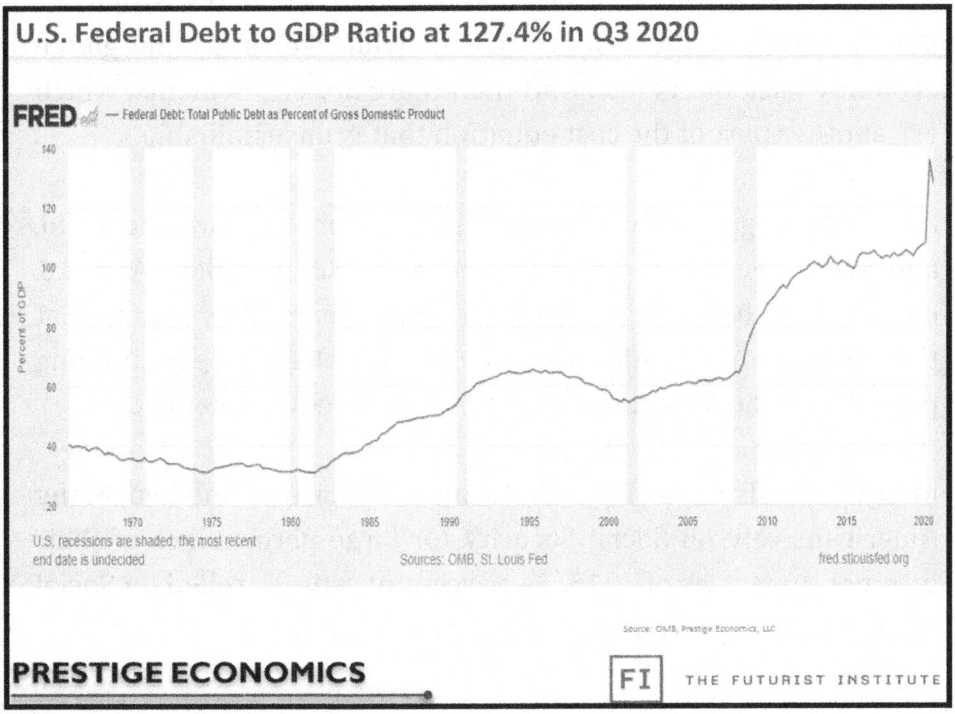

Although some analysts are quick to note that our debt-to-GDP ratio is lower than other countries, it is also important to note that the U.S. economy is the largest in the world. This means that rising U.S. debt levels could make it more difficult for the global economy to absorb U.S. debt issuances over time.

But hey! At least we're not alone. In fact, according to IMF data released in June 2020, the entire global economy has binged on debt in order to survive the current pandemic and recession. Now, both advanced economies and emerging market economies have seen the highest debt-to-GDP levels since the Second World War — or higher.[4] And these debt-to-GDP ratios are poised to rise even further.

But that's not even the worst news. The worst news is that we don't know how much more debt it will take to get the global economy back on its feet. And then there are entitlements, which are another part of the cost equation that is unsustainable.

A few years ago, the unfunded obligations for Medicare, Medicaid, and Social Security were estimated to be $200 trillion.[5] And U.S. entitlements have been underfunded for years. This means that the expenses to pay for entitlements have already been drawing partially on the operating budget of the U.S. government.

Plus, this is also a problem that cannot be easily remedied. Many Americans rely on Social Security for large percentages of their incomes. In fact, as of 2015, 34 percent of retirees relied on Social Security for 90 percent or more of their income.

In addition to this shocking number, in Figure 10-3 you can also see that an outright majority of all beneficiaries rely on Social Security for 50 percent or more of their incomes. Then there's also the demographic issue and the fact that baby boomers have been retiring *en masse* in recent years — something that could accelerate because of COVID.

This has contributed to a sharp rise in the number of new Social Security awards to retired workers, which you can see in Figure 10-4. In fact, the number of new annual awards for retired workers has doubled in the past 25 years, with the most recent year (2019) showing over 3 million new Social Security awards issued to retired workers.

Figure 10-3: Social Security Reliance[6]

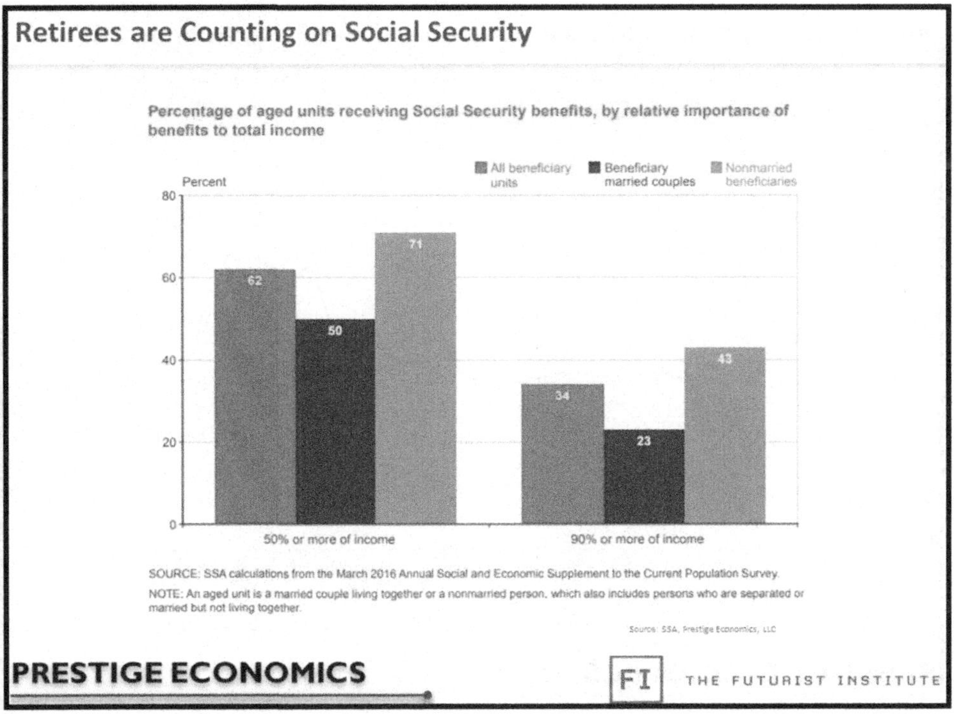

This graying of the American economy is something that contributes to the risks associated with the sustainability of the U.S. Social Security and entitlements systems. In fact, according to the report issued in July 2020 by the Social Security Administration, insolvency lies ahead.[7]

Furthermore, insolvency is expected to be reached in 2035, which is shown in Figure 10-5. This was already a path we were on before the COVID-19 pandemic, which should be clear from the fact that this SSA report was released in July 2020. It also means that this will still be a major threat after a COVID vaccine is widely available. In fact, this insolvency could happen more quickly than the July report warned.

Figure 10-4: Social Security Retirement Workers Rising[8]

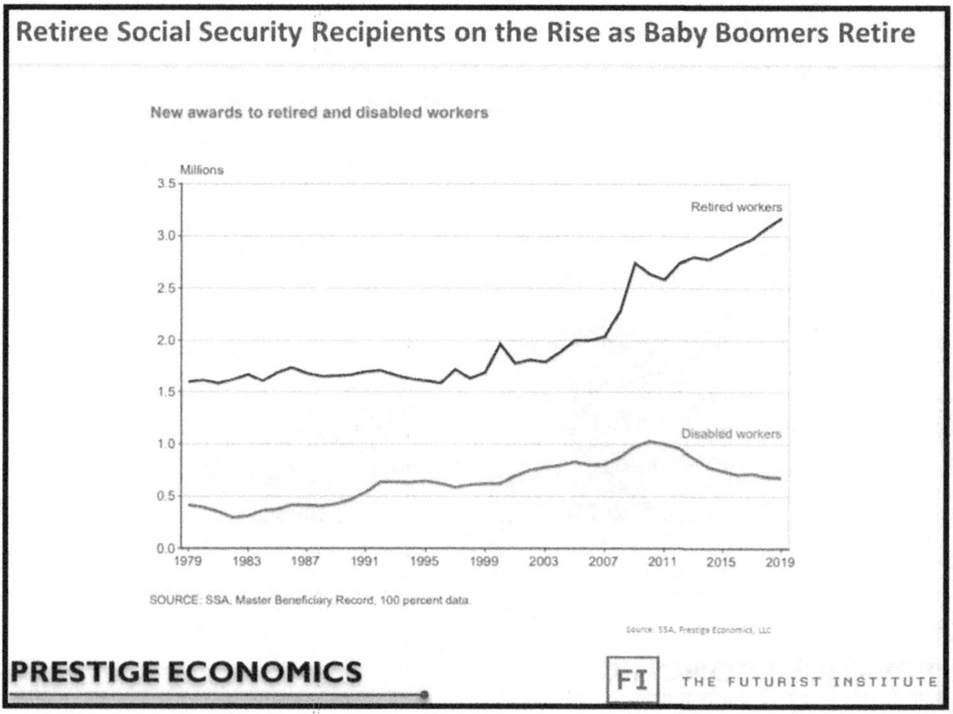

The exact verbiage from the Social Security Administration report about insolvency is as follows:

> Social Security is not sustainable over the long term at current benefit and tax rates. In 2010, the program paid more in benefits and expenses than it collected in taxes and other noninterest income, and the 2020 Trustees Report projects this pattern to continue for the next 75 years. The Trustees estimate that the combined OASI and DI trust fund reserves will be depleted by 2035. At that point, payroll taxes and other income will flow into the fund but will be sufficient to pay only about 79% of program costs. As reported in the 2020 Trustees Report, the projected shortfall over the next 75 years is 3.21% of taxable payroll.[9]

Figure 10-5: Social Security Insolvency Ahead[10]

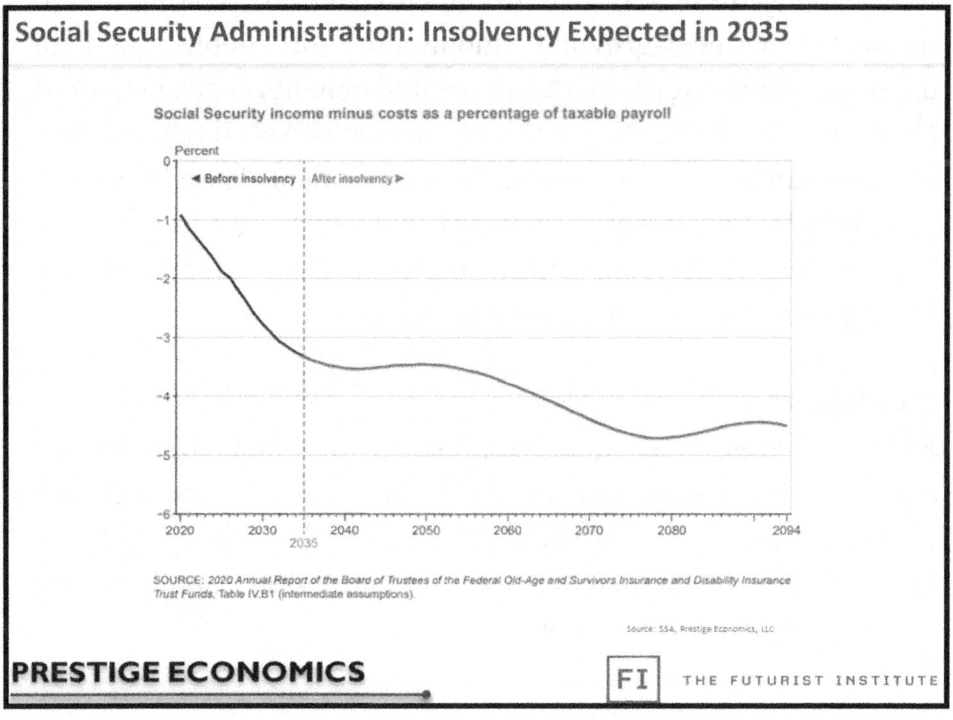

That is a pretty dire warning. But it could actually be worse for two big reasons: even more money is going out *and* less money is coming in than expected. And this is a dynamic that could be exacerbated in the COVID recovery period because unemployed people decide to retire when unemployment benefits run out.

In addition to rising outlays for more retirees who are awarded Social Security, the biggest entitlement risks to the economy and government finances may be Medicare and Medicaid costs. Those costs have been rising sharply and they are likely to be even more elevated in the current environment due to the asymmetric negative impacts of the COVID-19 pandemic on older people, like retirees. Vaccinating them first is also critical for Medicare costs.

Additionally, these are just the *expenses* we are talking about. Interest on ever-growing mountains of debt, sky-high unemployment rates with the associated benefits costs, increased Medicare and Medicaid costs, and the cost of Social Security are just one-half of the equation for the government. There's also the income side. Entitlements financing is likely to be soft for some time to come. After all, with a higher level of unemployment, fewer people will be paying payroll taxes.

Two Ways

Although some may take comfort in the fact that foreign exchange rates may remain stable because all countries are taking on mountains of debt, the future of global economic growth seems significantly less certain with rapidly rising amounts of government debt.

In fact, the entire situation reminds me of a discussion in Hemingway's novel *The Sun Also Rises.* At one point, a character is asked a question that seems apropos here: "How did you go bankrupt?"

His reply: "Two ways. Gradually, then suddenly."

When I think about the fact that it took the level of U.S. debt 205 years to reach $1 trillion but that the debt rose by more than $1 trillion per month in each month of the second quarter of 2020, it seems not unforeseeable that U.S. debt default risk is rising. It also likely means that further significant purchases of U.S. government debt by the Fed lie ahead. After all, someone needs to buy U.S. debt at big enough volumes to keep rates low.

Beyond the Nation

There is one final piece of the debt puzzle that we need to consider: state debt. It faces greater risks than federal debt. In fact, all government debt aside from national debt faces much greater risks ahead. After all, a national or federal government can print more money or direct its central bank to buy all its debt. For a city, county, or state, however, that's not an option.

Yes, central banks can buy muni debt. We have most certainly seen the Fed go down that road. But these debt issuers are at the mercy of national governments and central banks. And with imminent shortfalls in income, sales, and corporate tax revenues, these entities will be further stressed.

Additionally, if unemployment remains high, state government expenses could be higher — and a looming housing crisis could also set in, dampening potential property tax revenues. All of that does not bode well for muni debt, even if the Fed is enlisted to buy muni debt in ever-greater volumes in the years ahead.

Demographics

U.S. population growth has slowed sharply, and this demographic shift appears unstoppable. Plus, as birthrates have fallen, life expectancy has also risen. This compounds the funding shortfalls for entitlements. Worse still: No president, senator, or congressman can change U.S. demographics. This is bigger than one person. And a discussion of this was nowhere near the 2020 presidential election, as I predicted in *The Future After COVID*.

Population growth in the United States has fallen from annual rates of over 1.5 percent per year during the 1950s and early 1960s to just 0.7 percent since 2011.[11] Some of this slowing in population growth is due to a decline in the U.S. fertility rate. In general, fertility rates have been dropping globally, but according to demographer Jonathan Last, the U.S. fertility rate is still relatively high at 1.93.[12] However, even though the U.S. total fertility rate is relatively high compared to other industrialized nations, it is below the 2.1 percent "golden number" required to maintain a population, according to Last.[13]

This is a huge problem for maintaining entitlements. After all, the entitlement system worked really well in 1940, when there were 159.4 workers per beneficiary (Figure 10-6).

This contributes to an even more challenging situation, because that number fell to only 2.8 in 2013. Plus, it is likely to fall to 2 workers per beneficiary by 2040.[14]

Everything might be OK — if U.S. population growth were extremely robust. But it is not. Plus, the current administration is pushing hard to reduce illegal immigration to the United States. While this can have some benefits for society and the economy in some ways, it can also reduce the pace of population growth and lower the average U.S. birthrate.

Figure 10-6: Ratio of Workers to Social Security Beneficiaries[15]

Year	Covered Workers (in thousands)	Beneficiaries (in thousands)	Ratio
1940	35,390	222	159.4
1945	46,390	1,106	41.9
1950	48,280	2,930	16.5
1955	65,200	7,563	8.6
1960	72,530	14,262	5.1
1965	80,680	20,157	4.0
1970	93,090	25,186	3.7
1975	100,200	31,123	3.2
1980	113,656	35,118	3.2
1985	120,565	36,650	3.3
1990	133,672	39,470	3.4
1995	141,446	43,107	3.3
2000	155,295	45,166	3.4
2005	159,081	48,133	3.3
2010	156,725	53,398	2.9
2013	163,221	57,471	2.8

Furthermore, there are growing indications that the COVID-19 recession and pandemic could result in a baby bust, as noted in a July 2020 *Bloomberg* article by Peter Coy.[16]

The risk of debt losing value is significant. It means that those investors who hold fixed income could see their investments become worthless. This is a growing risk for muni debtholders, and it could become more apparent in 2021, as states and municipalities reel from tax revenue shortfalls, despite coming vaccination availability.

Of course, we could also see debt defaults by national and federal governments as a result of crippling COVID-19 debts, but printing money and directing central banks to buy the debt seems more likely. This means that even without explicit defaults, we could see more inflation, which could devalue fixed income payments.

Given high post-COVID debt levels — and the potential for them to rise even further — it is not inconceivable that we could also see negative interest rates in the United States at some point in future. Such a move could increase the relative value of existing fixed income interest rate products, but it would not provide positive yields on future fixed income investments. It could be quite the opposite if rates go negative.

As for the future of government services and the ability to support the populace through an economic crisis long after COVID, that potential will become diminished as the federal debt continues to rise as well.

It also means that hopes of universal basic income are very likely folly, because the government is already slated to struggle to pay for entitlements and the interest on its debt, let alone be able to cut endless new checks for people who may or may not be working.

There really seem to be only two ways forward for government expenses. Either interest rates will go negative, and countries (and other debtholders) will perversely earn interest on their debts. Or inflation will become rampant, and the debt will be devalued. In my opinion, since the demand side of the U.S. economy (and other advanced economies) is currently weak, negative rates appear to be more likely than consumer inflation — especially since they have already existed in the Eurozone for over six years.

Of course, governments could raise taxes to cover the shortfalls. But raising taxes is the kind of fiscal policy austerity that politicians and economists seek to avoid at all costs in the wake of a recession. After all, higher taxes would slow growth. And that could result in an even more uneven and checkered economic and business recovery in 2021.

CHAPTER 11

THE FUTURE OF FINANCE

For years, the field of finance has been undergoing the digital transformation that many industries got their first taste of during 2020 and the COVID-19 pandemic.

The reason? Finance is an industry that rapidly adopts new technologies, because technology has historically been an important lever of opportunity — and profits.

As we move into the new, new normal of 2021, the world has a lot more debt, central banks have bloated balance sheets, money is super cheap, and there are millions who may be unemployed for years to come. But the field of finance will continue experiencing perpetual disruption as technologies push to disintermediate cost, increase access, and improve user experience.

Plus, cryptocurrencies have experienced renewed interest and high prices in the face of expanded central bank balance sheets. But their existence could still prove ephemeral.

It's amazing that in our Rousseauian world of limited resources, the limit of anything — whether it be baseball cards in the 1980s, beanie babies in the 1990s, or Bitcoin in the past decade, scarcity often creates demand. At least that seems to be the case when money is cheap. Hype locusts of recent years have come back in spades, in part encouraged by central bank policies that threaten to further balloon an asset bubble that has expanded significantly since the last recession, despite a lack of consumer inflation.

The Hunt for Yield

The biggest challenge of recent years has been — and will be in the years ahead — identifying where people and companies are supposed to invest to generate returns. This is especially difficult in an environment of exceptionally low interest rates and commensurate low fixed-income yields.

Many assets already seem expensive: equities, real estate, private companies. Yet, consumables inflation is expected to remain low for years. The question of where to invest is going to be even more important, as the members of the U.S. Federal Reserve Open Market Committee — the FOMC — have indicated in their December 2020 forecasts that they expect to leave the main U.S. policy rate between 0 and 0.25 percent into at least 2024.

Plus, this is not just an issue for U.S. investors. It is a global problem. Yields have fallen almost everywhere. In Europe, the ECB had negative deposit rates for over six years between the fall of 2014 and 2020. And yields were still negative when this book went to print in December 2020. Plus, the ECB is likely to maintain negative deposit rates for the foreseeable future.

Where's the Beef — or at Least the Yield?

So, while technology threatens to push the field of finance into more leverage, fewer people, lower costs, greater access, and improved user experience, where is the yield? This reminds me of the 1980s commercial for Burger King, where several little old ladies examine a sad-looking hamburger and ask: "Where's the beef?"

For pensioners and retirees who wish to take less risk, this is going to be a challenge. Plus, while annuities had been gaining some new traction in recent years, interest rates near zero do not bode well for that asset class in the short run either.

These kinds of challenges threaten to remain in play long after COVID vaccines have been distributed, because the unemployment rate and level of joblessness are likely to remain high long after vaccines have become widely available. The dual mandate of the Fed to support full or maximum employment as well as price stability will have the FOMC keep rates low for years. And they even recently changed their inflation guidance to allow inflation to run hot for a time after COVID and exceed the Fed's 2 percent target for the PCE measure of inflation. But the FOMC's December 2020 median forecasts don't even reflect that U.S. year-on-year fourth quarter PCE inflation will exceed 2 percent through 2023, despite interest rates likely to remain between 0 and 0.25 percent and balance sheet expansions likely to continue. This is part of the reason for renewed interest in Bitcoin and other cryptocurrencies. It's what institutional investors call a positive return story. It means it makes money for investors.

Essentially, it doesn't matter if an asset or a company produces anything of value. As long as the stock price goes up, this becomes an investible asset. It is a self-fulfilling story. And it's one that hype locusts clamor for. It can also lead to dangerous financial market imbalances. It's also been a dynamic in equity markets with a relatively low number of investible assets against a backdrop of rising ETF purchases and cash sloshing around in need of a place to park.

Unreasonable IPO Dynamics

Another factor that has been driving yields lower and multiples higher is a greatly reduced number of investible assets. In fact, the number of publicly traded companies has fallen sharply — by almost half — in recent years, as can be seen in Figure 11-1. Furthermore, the number of publicly traded companies is likely to fall further, especially after COVID-19.

Increased M&A activity also seems likely to occur following the COVID-19 pandemic, shutdown, and recession for airlines, hotel chains, restaurant chains, automotive manufacturers, oil and gas companies. After all, many of these industries as well as other companies will be looking to spread their overhead across a larger organization that may be subject to reduced demand or downward price pressure for an extended period of time.

This trend of reduced investible equities leaves investors with fewer equity assets to chase. And it threatens to further push valuation multiples to unreasonable levels.

In addition to a forced concentration of buying and a reduced number of equities in the marketplace due to M&A activity, there has also been a significant slowdown in the number of new publicly traded companies that have been coming to market. In fact, there has been a significantly lower number of IPOs for many recent years, keeping the number of total listed public companies at relatively low historical levels. This dynamic can be seen in Figure 11-2.

Continued merger and acquisition (M&A) activity, ETF purchases, and a reduced supply of new companies through IPOs are a set of dynamics that threaten to reduce value opportunities for equity investors in the long run.

Figure 11-1: Number of Publicly Traded Companies[1]

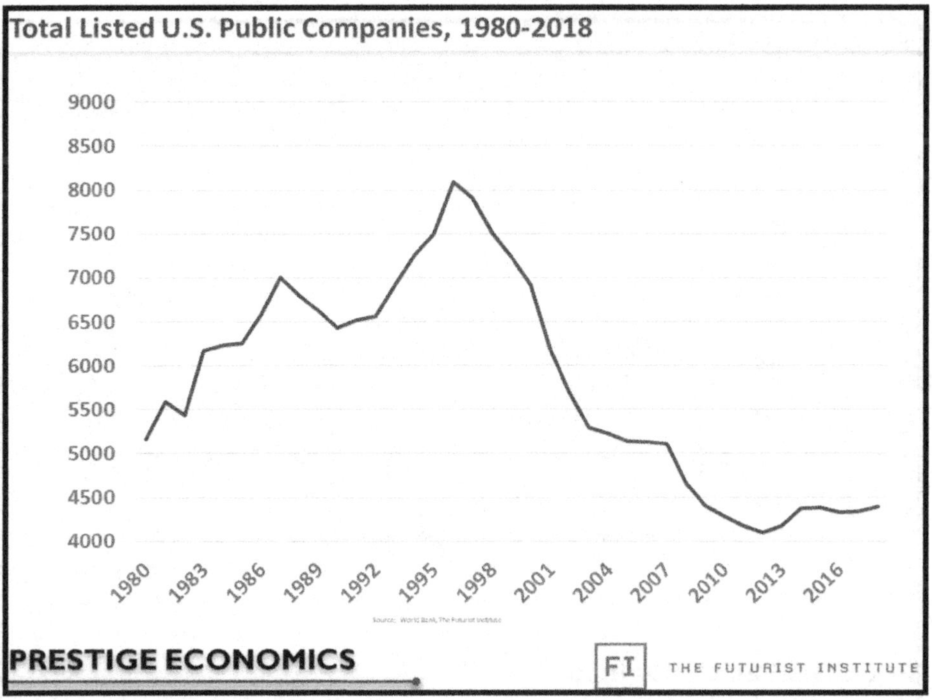

Plus, the number of IPOs in 2020 was low and the number in 2021 is also likely to be very low. As with other topics in this book, these dynamics are not new. And these dynamics are also highly likely to continue into the future of finance long after COVID-19 vaccines have been distributed. But that's not even the biggest risk in terms of chasing equity investment returns.

A bigger risk for the future of finance — especially in the near term — is that the percentage of IPOs with negative earnings was at 81 percent in 2018 — a figure that equaled the all-time-high percentage of negative earnings IPOs back in 1999, just before the tech bubble burst.[2]

Figure 11-2: Number of IPOs[3]

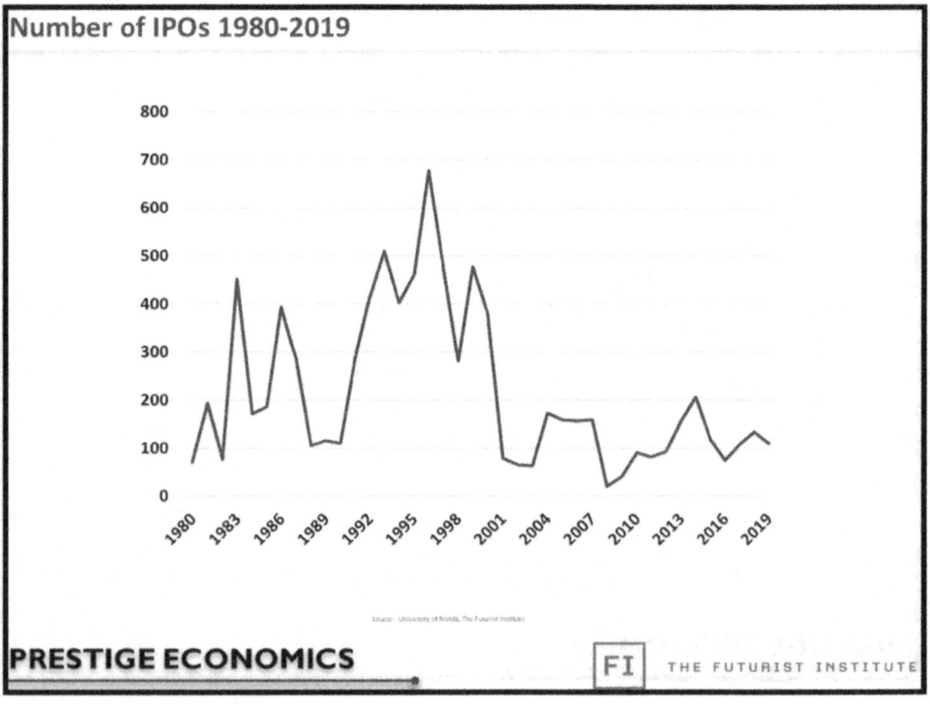

Historically, there has been a rising trend in IPOs with negative earnings, as you can see in Figure 11-3. But the rising trend of negative earnings IPOs during the most recent business cycle has been more drastic than in previous cycles.

Problematically, this could become part of the new, new normal in 2021 and beyond since there are now fewer public companies to invest in. Even if ETFs seek to diversify portfolios, with fewer companies to invest in, that becomes a challenge — and it may be creating additional demand for any kind of assets available, even if they have negative earnings. This kind of dynamic has likely been supporting some companies that fill a certain ESG plug or other requirement but may not necessarily be as relatively established from an income statement or operational standpoint.

Figure 11-3: Number of IPOs With Negative Earnings[4]

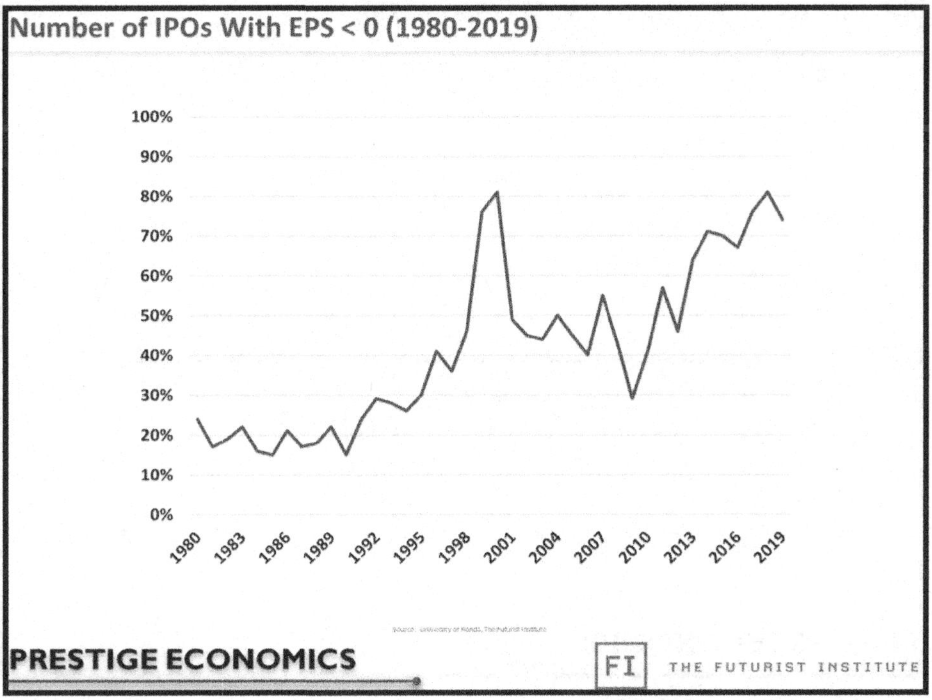

This may also partially explain why companies that have negative earnings have actually shown greater returns on their IPO days than companies with positive earnings. This dynamic can be seen in Figure 11-4.

Plus, the margin isn't a small one. In 2018, the average ratio of IPO-day returns for negative earnings companies was close to double that of companies with positive earnings. Although that figure fell in 2019, it remained high. Plus, a majority of IPO companies have shown negative earnings prior to IPO every year between 2013 and 2019, as shown in Figure 11-3.[5] As I noted earlier, some analysts and investors typify this surprising and backward dynamic as investing in a "positive return story."

Figure 11-4: Return Ratio of IPOs: Negative / Positive Earnings[6]

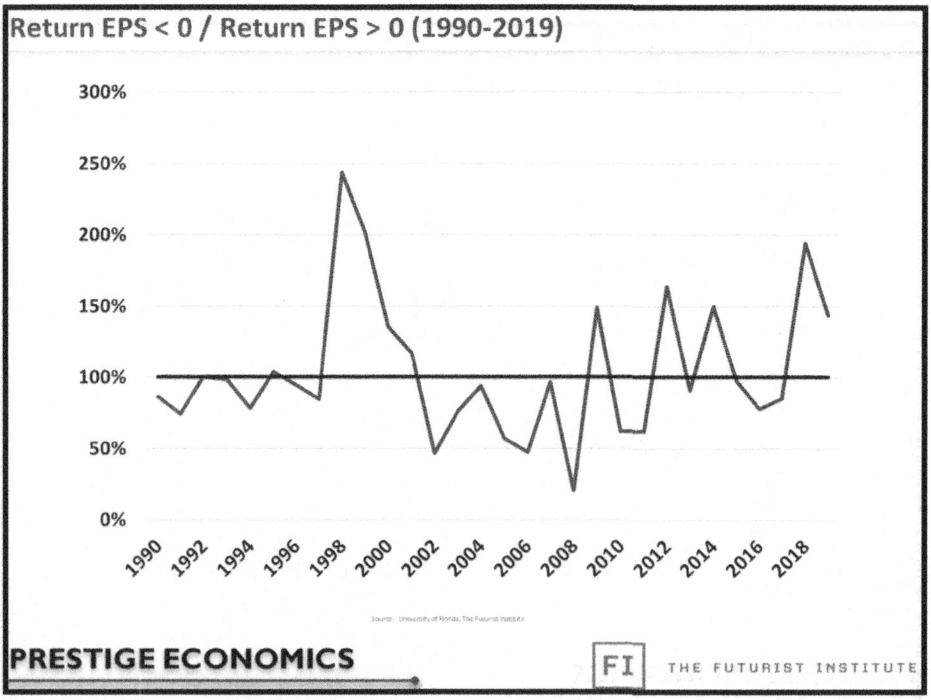

In other words, people make investments because the prices have always gone up. This dynamic begins with private funding, where different rounds of funding send valuations of pre-revenue and early-stage companies ever higher. From a seed round to Series A and beyond, all the way until IPO, companies with negative earnings see their valuations rise — despite losing money.

Although the future of finance is likely to continue to see the majority of IPO companies have negative earnings most years, the current percentage is near historic highs — and I have warned that we could be in for a correction in the event of a downturn, like the current COVID-19 recession.

For now, however, investors bemoan the lack of investible assets as the globally coordinated central bank quantitative easing party continues to introduce excess cash and liquidity into markets lacking a potential for returns.

The result has been a push into assets that may have inflated valuations. Furthermore, the problems of excess liquidity, lack of investible assets, and FOMO have also crept into the private investing space as well. In fact, I have personally seen deals that would boggle the mind of the average person with any basic understanding of business and logic.

Looking at the future of finance, I expect dynamics in public markets could continue to bleed over into private markets.

While quantitative easing, which I discuss in Chapter 9, was an effective way to stimulate a global economy on the brink of disaster during the Great Recession and the COVID-19 recession, it has also contributed to excess capital-seeking returns.

The quick moves to enact unprecedented accommodation by the Fed, the ECB, and other central banks helped buoy equity markets during the worst of the COVID-19 pandemic, but it has also made investments generally expensive again. Furthermore, a persistent period of monetary accommodation, coupled with expectations that the Fed and other central banks will always be the buyer of last resort for debt, incentivized some bad behaviors after the Great Recession and the global financial crisis.

But, as I noted in *The Future After COVID*, it could incentivize even worse behaviors after the COVID-19 pandemic. Indeed, we are beginning to see this kind of dynamic. Evaluating this situation as a futurist, it is disconcerting that historically, there has been a piper to pay when such dynamics continue for too long.

This means that even though equity markets were relatively quick to show a V-shaped recovery in the immediate wake of the COVID-19 economic shutdown, there are still risks ahead. The sustainability of a system that is now perceived to be bulletproof and unable to fail could make the hunt for yield even more challenging in the decade ahead. After all, financial investments offer tradeoffs of return for risks.

But if all assets are now riskless, will assets also be returnless?

Mixed Dynamics

For investors, the future of finance could be one of increasing risk if you want any hope of return. Essentially, central banks have crowded out buyers of relatively low-risk, stable-return fixed income investments. This means that investors from retirees and day traders to government pension fund managers and non-profit endowments are crowding into equities as one of the only places to achieve yield.

Against these elevated risks in the world of finance, there is also going to be a persistent push to reduce costs. Still, no matter how much you reduce costs — or how pleasant you make the user experience — by leveraging technology, if risk-return ratios are out of whack, there could be problems ahead. It may not be in 2021, but there are certainly rising risks, especially if we consider the P/E ratios of many publicly traded companies.

In other words, despite the wonderous benefits of technology in the world of finance, risks could continue rising for years because of these dynamics exacerbated by the COVID-19 pandemic — even after COVID vaccines have been distributed at scale.

CHAPTER 12

THE FUTURE OF REAL ESTATE

With the onset of COVID, real estate has been significantly impacted. I am not surprised. In *The Future After COVID,* I noted that real estate would be an industry with the potential to be significantly impacted by the COVID-19 pandemic and its aftershocks.

The push to remote work, the rapid acceleration of e-commerce, and the mandated shutdown of leisure and tourism industries had myriad impacts. In *The Future After COVID,* I noted that there were six high potential ways for dynamics to play out. Five of these have played out as I predicted:

— **Reduced commercial office demand.**

— **Reduced retail space demand.**

— **High risks to real estate in tourism-centric areas.**

— **Likely tradeoff preference for space vs. proximity.**

— **More demand for warehouse and distribution centers.**

There is also one way things have not played out as expected:

— **Risk of housing oversupply and price declines.**

The reason I acknowledged a risk of housing price declines was because of the exceptional weakness in the labor market. Prices have not declined broadly, because eviction and foreclosure holidays, as well as Fed balance sheet expansions, have kept the millions of jobless in their homes — and it kept the housing market afloat. But there are still risks.

Despite these 2020 dynamics, there are still risks to parts of the housing market after the COVID vaccine — in 2021 and beyond. After all, the labor market lags behind financial markets and the economy. As I noted in *The Future After COVID*, if jobs do not come back and businesses fail, there will be evictions, there will be bankruptcies, there will be mortgage foreclosures, and even though the government can help to bridge some of the lending and some of the gaps to prevent widespread national mortgage failures, there are going to be some areas where homeowners default on their mortgages. Furthermore, no matter how quickly vaccination occurs at scale, jobs will not return instantaneously.

Looking ahead at the other five dynamics I noted in *The Future After COVID*, some important developments seem likely in 2021 and beyond.

Commercial Office Demand

In 2020, many companies broke leases or put their commercial office expansions on hold. In 2021, companies may begin to bring some of their workforce back to the office. But the model — befitting of the new, new normal dynamics of 2021 — is likely to be a hybrid compared to 2019 and 2020.

If people continue to work remotely at levels anywhere near those seen during the COVID-19 pandemic, we will likely find that we have far too much commercial real estate for offices — especially expensive real estate in areas with high urban density. As I noted in *The Future After COVID*, some high-end commercial real estate could be converted into hotels, condos, and mixed-use buildings.

In the year ahead, when profitability and cost-cutting will be critical for many companies, those companies may not just be reducing current commercial office space needs. There is also likely to be an abatement of future office space demand as many workers continue to operate remotely and use their offices more as flexspace operating environments, the way some McKinsey offices have operated for almost a decade and a half.

Reduced Retail Space Demand

In *The Future After COVID*, I also wrote about my expectations that retail real estate would be significantly impacted by the pandemic. The trend of e-commerce has been rising for some time, and it accelerated massively in 2020 — especially in the second quarter, when it became economically critical during peak shutdowns of the COVID-19 pandemic.

As a result of COVID-19, a lot of small mom-and-pop businesses, family restaurants, small service businesses, and small retail shops have completely shut down and went out of business in 2020. For larger brick and mortar retailers, there are also risks. In general, e-commerce is likely to remain very strong — and continue rising over time.

This means that department store chains lacking a massive e-commerce presence, and which have been under pressure for years, have also hit significant financial tipping points in 2020. Many of them may not recover.

In 2021, the surviving brick and mortar retail businesses could see some recovery, but physical retail and restaurant industries are likely to remain decimated for some time. After all, banks and investors could be very reluctant to pour money into these industries for some time.

As of 23 November 2020, retail and restaurant bankruptcies of 2020 have included some high-profile companies, including:
— Ascena Retail (Lane Bryant, Ann Taylor)
— Brooks Brothers
— CEC Entertainment (Chuck E. Cheese)
— Century 21 Stores
— GNC
— Guitar Center
— J.C. Penney
— J. Crew
— Lord & Taylor
— Neiman Marcus
— Pier 1
— Stein Mart
— Tailored Brands (Men's Wearhouse, Jos A. Bank)[1]

Other companies will also end up in bankruptcy because of COVID. And while many could emerge from bankruptcy protection in solid positions, many may never fully recovery.

Other Service Businesses

The pandemic has had a particularly notorious impact on service businesses because these are high-touch businesses like massage therapy, nail salons, and hair salons. Some of these businesses are quite small and they are struggling to survive.

Real Estate in Tourism-Centric Areas

In *The Future After COVID*, I noted my concerns about real estate in tourist-centric cities like Las Vegas, Orlando, and New Orleans. Casinos and theme parks were forcibly shut down during COVID, because high tourist areas presented such high risks of COVID-19 exposure. But many of these have since reopened.

Despite staccato and limited reopenings in 2020, as well as prospects of more widespread and complete reopenings in 2021, tourism dollars could still remain tight with so many unemployed people. Disposable incomes might be slow to recover for many, and if that's the case, we could likely see major tourism be slower to recover than other forms of entertainment and leisure. After all, people are more likely to eat at a local restaurant or drink at a local bar than fly to Disney World immediately after they get their vaccinations.

Tradeoff for Space vs. Proximity

Following the outbreak of COVID, I expected an increase in the preference for residential real estate *space* over a close *proximity to work*. I noted in *The Future After COVID* that we might see home buyers demonstrate a preference for larger, suburban homes over condos close to urban centers.

Indeed, we have seen this dynamic play out. Real estate prices in suburbs, exurbs, and rural areas have risen sharply, while some urban real estate prices have stagnated or fallen. Rents in New York City and San Francisco, for example, have fallen sharply.

Many people were predicting greater trends in U.S. urbanization for most of the past 15 years or so. But the remote work experience has changed people's perceptions about the value of space. After all, people are working remotely with their entire families in close quarters — rather than going to an office regularly — population-level preferences are likely to shift toward more space. This is especially true for families that have adults who have been working remotely alongside their kids who have been going to school online.

During the pandemic period there has also been elevated interest in the United States in smaller cities, like Austin and Charlotte over Silicon Valley and New York. This is not a new trend, and it is not likely to end with widespread COVID vaccine distribution. After all, remote work and remote education are unlikely to cease with vaccinations. As such, the arbitrage value of real estate space will continue to close as remote work remains high and continues rising.

Warehouse and Distribution Centers

The final dynamic I noted of import for the future of real estate in *The Future After COVID* was the likely increase in demand for warehouse and distribution center square footage. The critical nature of the supply chain and the importance of e-commerce during the pandemic period have driven this demand further.

Plus, e-commerce and supply chain have been growing in importance for many years. And that dynamic is likely to continue on trend, even if there is some reversal in e-commerce as a percent of retail in coming quarters compared to the peak percentages in the second quarter of 2020.

Looking Ahead

Some real estate dynamics changed against a backdrop of the COVID-19 pandemic, especially the downside impact on commercial real estate markets and some urban rental markets. But some new dynamics have also been revealed, like the remote-work-induced preference for residential real estate space over proximity to work as well as the accelerated need for more warehouse and distribution center square footage.

While COVID-19 has impacted real estate markets in a number of ways, there will also be new changes with the rollout of a vaccine at scale. In the medium term, interest rates are likely to remain low for a number of years. This is supportive for real estate valuations across the board. Additionally, elevated levels of remote work and e-commerce will be critical drivers for real estate markets in the short term as well as the long term.

CHAPTER 13

THE FUTURE OF SUPPLY CHAIN

Paper product and fresh food shortages in the United States resulting from the COVID-19 pandemic came as a surprise to many people. Because I was born in the 1970s, I recall images and stories of life in the former Soviet Union, where people were unable to get toilet paper and food. The fact that this was America during the COVID-19 pandemic of 2020 was also shocking to me.

These dynamics will cast a long shadow over people's careers and lives in a material way long after a COVID vaccine is widely available. Even now, many people have already come to recognize the importance of the U.S. supply chain, e-commerce, material handling industries, and last mile challenges.

The vulnerability of U.S. and global supply chains have been revealed during the COVID-19 pandemic. Although much has been done to address these issues, there will still be a lasting impact in the post-pandemic period, long after vaccines are widely available. It is also now clear to many that disruptions in the supply chain can occur anywhere in the global economy and negative impacts are exacerbated when inventories are lean.

In short, the long-term trends of simultaneously spreading out supply chains while maintaining low inventories can be a recipe for disaster. Although it largely did not come to disaster during the COVID-19 pandemic of 2020, it was close.

Where our goods come from, how they get to us, and when they get to us are critical issues that people can no longer ignore. And it may be somewhat of a near-term challenge to instill great confidence in a supply chain that is spread very thin — and spread globally.

Beyond the overall awareness of supply chain risks, and people's personal experiences with food and paper product shortages, the perceived importance of medical supply and medical device supply chains has also increased. This is something I expected and wrote about in *The Future After COVID*, and it is a topic that is front-and-center as pharmaceutical companies rush to deliver vaccines at scale to stem the spread and public health impact of the COVID virus.

In general, we have long taken supply chains for granted. And this is doubly true of medical devices, supplies, and personal protection equipment (PPE). In the future, the medical supply chain is likely to be recognized as more critical. And I continue to expect that U.S. government policies are likely to be enacted to reduce medical supply chain risks. This is also something I noted in *The Future After COVID*, and I expect that the focus on supply chain risks will continue under a Biden administration in the United States, as it will in many other countries.

Overall, the supply chain of the U.S. economy is something that almost everyone now more clearly understands. The words "supply chain" were barely uttered in business schools in the 1990s and early 2000s. But now, supply chain needs to be a top-of-mind topic for every executive, business leader, consultant, and consumer. Maybe the perceived importance of supply chain will fade over time. But as long as people remember the COVID-19 pandemic, it is likely to leave a lasting imprint.

The Importance of Supply Chain for Stability

Stability of the economy depends very much on goods and services getting to where they are needed. And a disruption of those goods presents not just risks to the economy but actual risks to people and potentially to national security. Fortunately, the United States is a net exporter of food. But other countries are not so lucky.

When talking about people and how they live, Berthold Brecht, the playwright of *The Threepenny Opera*, wrote the line "Erst kommt das Fressen, dann kommt die Moral." It translates roughly to "First comes the feeding, then come the morals." In other words, if people do not have access to food, this could destabilize an entire society.

Fortunately, that did not happen during the pandemic, but it could still skew people's perceptions and behaviors for a long time to come. After all, long after vaccines have been widely distributed, people may be less likely to run down their own "at-home" inventories of food, paper products, cleaning products, and other goods.

There's an old nursery rhyme about Mother Hubbard and her cupboard. As you may recall, her cupboard was bare.

Because people have been able to get goods with relative ease for some time, people did not worry about what was in their cupboard. After all, you don't mind if you have a bare cupboard if you can get whatever food you need to your house in under 20 minutes.

But in the COVID-19 pandemic situation, people found supply chains disrupted while their cupboards were also bare. This problem became exacerbated by the risk that people might need to stay in their homes for extended periods. People had to restock their cupboards rapidly because they had low in-home inventories and they expected a rise in demand as alternative sources of food (e.g., restaurants) became less viable. This is what contributed to the bullwhip effect of demand for food and paper products.

The food supply chain as well as the supply chain for paper products was shifted from being consumed at home *and away* to solely being consumed at home. And there were no reserve foodstuffs; there was no back of house. Mother Hubbard's cupboard was bare everywhere, and people had to react quickly. In the future, consumers, policymakers, and governments will want more secure supply chains. But it also seems likely that consumers will take some of the risk into their own hands and prevent their cupboards from going bare by keeping more inventory at home long after COVID vaccines are available, just in case some future unknown crisis hits.

Expectations Summary

COVID-19 revealed a few problems for supply chain that are likely to remain top of mind. We also know that the economic and policy drivers are likely to be impacted by basic individual and societal fundamentals, which are usually solid grounding for futurist scenarios and expectations.

The most important fundamentals for the future of supply chain after a COVID vaccine is widely available include:

— People always want to have access to food.

— People want to always feel secure in their access to medical care.

— Society only functions if people have access to food and medical care in a way that they feel secure in it.

This is why I believe we could see additional financial or regulatory incentives from the policymakers and governments to shore up supply chains in the future — and to make them more resilient in the face of a crisis. After all, by any stretch of the imagination, COVID-19 appears to have been a very close call for supply chain security.

CHAPTER 14

THE FUTURE OF INTERNATIONAL RELATIONS

The COVID-19 pandemic did no favors for the state of global affairs, and specifically the already tense relationship between the United States and China. President Trump's insistence on still calling COVID-19 "the Chinese virus" underscored high levels of tension just below the surface. And the struggle between China and the United States for global economic, political, and military hegemony was not softened by recent developments.

In fact, it appears as if COVID-19 may sow more seeds of discontent along with increased tension between the two competing nations. After all, while COVID originated in China, the Chinese economy was one of the first to quash the virus and return to normal, with Chinese manufacturing rising to the strongest monthly increase in November 2020 since November 2010 — a decade earlier.

The International Monetary Fund has also been consistently forecasting that the only major economy to grow in 2020 would be China. This means that China has thrived in this environment, unlike the United States and many other countries.

The so-called Great Power Competition between the United States, China, and Russia has been playing out on the world stage for years — and risks have been rising. Looking to 2021 and beyond, it seems likely that the approach the incoming Biden administration takes toward these other two global powers will be more important than the distribution of a COVID vaccine at scale. Of course, this does not mean a Biden administration will completely change policies toward China — and the new administration seems likely to increase the pressure it puts on Russia.

While the relationship between the United States and China was already strained by a trade war that started in early 2018 and continued throughout 2018 and 2019. Even now, some of the key U.S. tariffs still remain in place, and they may not vanish overnight under a Democratic Biden administration either. As for the relationship between the United States and Russia, it seems likely that the Biden administration will push for closer ties with NATO and U.S. allies in Europe. This could escalate tensions with Russia.

There are some important lessons from the COVID-19 pandemic for the United States. Perhaps most importantly, there is the fact that this was a global pandemic. If COVID-19 had only — or primarily — impacted one country, like the United States, that could be quite a bit more challenging for that country's geopolitical standing. Such a risk could in turn present stabilization problems, and it could engender asymmetric negative geopolitical impacts on a country that is solely afflicted.

This highlights the future risks, potential, and power of biological warfare. On a global scale, risks to medical supply chains have also been exposed as a real risk for the safety of the American people. Even under a Biden administration, there may be medical device, medical supply, or PPE tariffs put in place to prevent the supply chain problems and economic chaos that the COVID-19 pandemic outbreak created — and exposed. After all, next time we might be in an even worse situation.

In the long run, tensions between the United States and China are likely to lead to de facto bifurcated global supply chains at the global level. While this could disrupt the Chinese technology ecosystem, the pandemic has resulted in sharply stronger Chinese manufacturing activity. This means that while U.S. policies may be designed to disrupt Chinese global manufacturing hegemony, China's pole position has only become more solidified by the pandemic — if IMF and recent PMI data are to be believed. Plus, China could make further relative gains in coming months as the United States and other countries lurch forward with staccato, patchwork COVID vaccine distribution toward a post-pandemic world.

Although recent and current tensions seem unlikely to put the United States and China on a destined-for-war footing, there is significant potential for further erosion in relations between the two countries that could lead to increasingly risky outcomes, including the potential for proxy wars in the future operating environments of 2030 and beyond.

CHAPTER 15

THE FUTURE OF NATIONAL SECURITY

The COVID-19 pandemic has opened the aperture in many ways, revealing open secrets and overlooked risks — for our society, our economy, and our security. For national security, the COVID-19 pandemic has exposed the importance of being vigilant with our borders. And it highlights the risks of overly thin supply chains, as well as the potential downside of being dependent on the global supply chain for critical goods like medical supplies, medical devices, basic necessities, and personal protective equipment (PPE), including gloves and masks.

Supply Chain as a National Security Risk

Overly thin levels of inventory as well as long, lean supply chains present risks to national security in the United States. Most corporate entities have run their supply chain inventories exceptionally low.

This thin level of supply — this lean level of inventory — produces risks, as does the geographic distance between trade partners, which introduces a lag time between which goods can be ordered and received.

As we have seen, this has been critical for medical devices and PPE. Trade risks were already a national security issue related to metals for materiel in the United States, which was the critical driving factor behind the implementation of the U.S. Section 232 tariffs on aluminum and steel. Additionally, the Section 301 tariffs highlighted risks to U.S. national security from Chinese threats to U.S. intellectual property.

Due to the crisis of COVID-19, PPE and medical devices are on the hot list too. As a result, the U.S.-China trade war is likely to heat up from current levels, because risks that were previously considered somewhat fringe are now likely to be accepted as national security risks on a bi-partisan basis. In other words, both Democrats and Republicans may now come to see having critical goods, including metals for materiel, PPE for healthcare, foodstuffs, and paper products as important for maintaining civil order. In some respects, maintaining a secure supply chain is inherently difficult when the goods come from a country that is too far away for the goods to arrive quickly.

That realization is now going to receive more airtime — more talking space and room for discourse. So, we may very well see in the near term that both Democrats and Republicans will trip over themselves in order to support the idea of securing supply chains for critical medical devices, PPE equipment, foodstuffs, paper products, and pharmaceuticals to prevent the loss of American lives in the future — and to minimize the potential disruption or devastation of the American economy in the event of another pandemic.

After all, one of the big reasons we needed to slow the spread of COVID-19 in the United States and "flatten the curve" was because there were not enough doctors, nurses, hospitals, ventilators, gloves, masks, or other equipment to handle a pandemic. If we could at least get the equipment piece under control, we may better protect the economy and the American people in the future.

Risk of Exploitation

The COVID-19 outbreak has also revealed that the United States can be exploited in a pandemic-type event. Furthermore, if the United States had been solely targeted, it would have significantly threatened national security. Alternatively, if the United States could have been deceived into believing that flattening the curve could be beneficial — even when it was not — that, too, could have presented a risk.

This experience also reveals the vulnerability of the American public to media and social media messaging and potential manipulation around topics that may or may not be valid. These are additional risks that could be presented by psychological operations and the adoption of subjectivist truth.

In short, the United States, from a national security perspective, may very well wish to more firmly shore up technology, healthcare, medical, food, consumables, and other supply chains in order to ensure maximum stability over time. Additionally, COVID-19 has inadvertently revealed that a pandemic-level bio attack on the United States could be beyond economically devastating.

If adversaries of the United States were to implement such an attack, they could pair such action with social media and traditional media disinformation for maximum disruption, political destabilization, and economic devastation in a way that could completely destabilize the United States as an entity for at least a brief period of time.

These kinds of risks may sound extreme.

But this was an extreme situation. And it will not be the last time our economy or nation face extreme risks. After all, we have also seen foodstuffs run low, we have seen shortages of paper products, and we have seen the American populace need to quarantine at home.

On the policy side, the Fed massively expanded its balance sheet as I noted in Chapter 9. Meanwhile, fiscal policy stimulus in the Coronavirus Air, Relief, and Economic Security (CARES) Act and more. These policy actions alone underscore the real costs of being unprepared — or even just underprepared.

If we were to consider the cost of having just produced more ventilators, masks, and gloves — and if we had simply educated more doctors and nurses — those costs would have likely been infinitesimal compared to the potentially multi-trillion-dollar level of fiscal and monetary policy support enacted to prevent a collapse of the entire U.S. economy.

"NOISE" Framework

When The Futurist Institute considered the importance of the COVID-19 pandemic for national security, I created something I call the NOISE framework, which examines some of the most important factors that contribute to national security and political stability.

Here is the NOISE framework and the five critical factors that contribute to national security and political stability:

Necessities — *Food, Water, Power, Shelter, Safety*

Occupations — *Jobs, Vocations, and Hobbies*

Information — *Access to Accurate, Complete Information*

Systems — *Financial, Healthcare, Transportation, Education*

External — *International Relations, Military, Supply Chain, Trade*

As you can see, first come the necessities, which I have included as food, water, shelter, and safety. If people do not have these, then there are easily risks of political instability.

If these factors are stable, then a country or economy is likely to remain on an even keel. In the most recent experience of the COVID-19 pandemic, concerns about safety and food emerged. This is critical, as is maintaining the U.S. supply chain and basic services as well as utilities (like power and water).

Manufacturing and National Security

In addition to broad-based national security risks that have been revealed by the COVID-19 pandemic, there are also risks associated with disruptions to national security vendors.

Some companies that produce goods for national security uses (i.e., airplane manufacturers and airplane parts manufacturers) could find themselves suffering from challenging economic and business conditions as part of the slowdown in air travel. While this is a second-order impact of the COVID-19 pandemic, it is a critical issue of primary importance for national security entities.

Looking ahead, national security organizations will need to more aggressively monitor economic and business risks of critical vendors. This is just as true for large publicly traded vendors as it is for startups that provide essential materiel to the defense industry.

How the government addresses these risks in the future is up for debate and unclear, but one thing seems certain: Even if these kind of national security vendors may not be too big to fail, they may be *too important* to fail.

Four Scenarios for National Security

In order to help the United States Pentagon think strategically about how the aftermath of the COVID-19 pandemic could impact national security, I was tasked with creating four potential scenarios for the U.S. Air Force Warfighting Integrating Capability (AFWIC) unit. My scenarios became part of an unclassified Pentagon report that was released in June 2020.

One of the most critical elements of these scenarios is how the future of international relations between China and the United States could change — and how it may create ripple effects in global supply chains, the global economy, and great power competition.

Here is a link to the full unclassified Pentagon report:
http://www.futuristinstitute.org/pentagon-report/

My scenarios highlight how national security, trade, and supply chains may be affected by the COVID-19 pandemic. There is more color included in this chapter than in my section of the report. Also, unlike in the Pentagon report, in this book I discuss which scenarios and potential futures seem most and least likely below.

Potential Future A: The Hollywood Ending

The international competition between the United States and China becomes deprioritized. The competition between the United States and China gives way to international peace and collaboration on multiple levels, including scientific research, supply chains, and vaccine development. All concerns fade about initial COVID-19 data transparency, long-term international competition, and the Chinese push toward economic, political, and military hegemony over Asia.

This would represent a truly transformational-type outcome, given decades-long trends in place and elevated concerns in the COVID-19 pandemic era. It is also the least likely scenario as seen from December 2020.

Potential Future B: Trade War Détente

The hopes for an end to the U.S.-China trade war that were high at the end of 2019 are revived in the wake of the COVID-19 pandemic. Some trade restrictions remain in place, but some are dialed back. If there are additional future trade restrictions, these may be quite limited.

The sense of economic symbiosis increases, as opposed to a sense of economic competition. This is a challenging dynamic, as it necessitates an important precondition of reduced competing counterparty rhetoric.

This scenario also requires a reduction of national security concerns about forced technology transfers, supply chain security, and a propensity for military conflict and influence in the South China Sea. Although more likely than Future A, this also seems less likely than Future C.

Potential Future C: The Best of Frenemies

The competition between the United States and China ramps up over a period of quarters or years. The U.S.-China trade war enters an accelerated phase, in which trade restrictions imposed by the Section 232 and Section 301 tariffs are augmented by additional tariffs on medical devices and materials.

Competition also manifests itself in additional supply chain restrictions, prioritizing U.S. consumption of goods from the United States and the USMCA region. Manufacturing increases as a percent of the U.S. economy's GDP. Not many manufacturing jobs may be added, however, as robots do much of the work.

Some technology companies as well as other priority firms are forced to implement bifurcated supply chains to serve both the United States and China. Relations between Washington and Beijing take on an uneasy "frenemy dynamic," punctuated by alternating moments of symbiosis and elevated competition.

For the record, this is the scenario that I see as most likely at the moment. Although there are emerging reasons to believe that there could be an even worse scenario, which is below.

Potential Future D: Destined for Proxy War?

The fallout from COVID-19 leads to an almost immediate direct conflict between the United States and China. A Congressional inquiry into COVID-19 data sharing as well as exposed medical device supply chains leads to significant trade restrictions and geopolitical backlash against China.

Supranational organizations like NATO and the United Nations are pulled into the conflict. Feeling threatened, the Chinese government ramps up its ground game in countries suffering from COVID-19 to shore up support for its strategic objectives. As disinformation campaigns, soft policy diplomacy, and directed influence funding accelerate, proxy wars become almost inevitable. The result would be armed conflicts between proxies.

While much of Asia could be embroiled in such proxy wars, one of the greatest risks would be posed by a proxy war in India. This is not just because of India's size, but because it could be seen as furtive ground for ideological and social conflict, given its heterogeneous linguistic, ethnic, and religious composition.

Reviewing Scenarios

Potential Future C seems to be the most likely future scenario as seen from my vantage point in December 2020. But the outlook and risks can change rapidly, as we have seen during the COVID-19 pandemic itself.

This is why my most dire scenario could understate future negative realities, just as my most optimistic scenario may understate positive potential for the future. It is also the reason why I was tasked with creating scenarios for four different potential futures.

Listening for NOISE

As for the minimization of risk to national security, it will be critical to control for and support those elements in the NOISE framework that represent the fundamental building blocks of stability. That's where the risks are, and only if those levers of stability are firmly in place can the risks to national security and society be minimized. This is just as true now as it was in the first days of COVID. And it will remain true whenever the next public health crisis rears its ugly head.

CHAPTER 16

THE FUTURE OF TRAVEL AND LEISURE

Travel and leisure industries were devastated by COVID-19. Looking ahead to a vaccinated post-pandemic 2021, there is optimism. But vaccinations, herd immunity, and a return to pre-COVID joblessness will take time. This means that there could very well be continued aftershocks for the world of tourism and travel for years to come.

The future of travel and leisure hinges very much on a question I noted in *The Future After COVID*: How quickly after COVID-19 passes will you be willing to travel to Vegas and go to a casino — or go to Orlando and visit a theme park?

This is a question that I predicted would drastically impact travel in 2020 and 2021. It is a question with mixed answers. After all, some people have already returned to theme parks and casinos. But they are relatively small in number. As we look to the year ahead, that number is likely to rise significantly — especially by the end of 2021, as people emerge from their reduced travel activities to go out into the world.

Risks to Tourist Centers and Related Businesses

Some of the biggest negative risks and impacts from COVID-19 this year have been on tourism-centric regions. I think of the joblessness rates in places like Las Vegas, Orlando, New Orleans, and Hawai'i. And the negative impacts could have significant economic shockwaves for other cities with a lot of other tourism and conference demand, like New York City, Austin, Houston, San Diego, Asheville, and many others.

Part of the negative economic consequences of reduced travel and tourism was engendered by government regulations, mandates, and policies designed to help people avoid crowds. But there has also been a reduction in disposable income for the millions of newly unemployed Americans that could also have negative impacts on tourism and tourist centers in the year ahead — and beyond.

This loss of business has directly impacted the workers and companies in the tourism industry. And although these jobs are likely to return in the year ahead, the return could be slow and uneven. Plus, workers in tourist industries may take a once burned, twice shy approach to the sector, seeking out training and education to find new employment opportunities in more stable sectors.

As we look at 2021 and 2022, I expect that some individual travel and tourism may remain depressed, because of the financial impacts. But business travel — especially to high-value conferences — could actually increase in 2021 and more significantly in 2022.

Changes in Business Travel

As road-weary business travelers were forced into a moratorium on travel and a perpetual staycation in 2020, some have been plagued by a sense of wanderlust, but many will likely seek out ways to engage in more efficient business travel. This means that even after COVID vaccine distribution, there will still be many Skype or Zoom meetings, hybrid conferences, and a preference for in-person conferences over multi-week roadshows from client to client or customer to customer.

This is just one of the ways that the COVID-19 impacts could be felt for years to come.

Impacts of a Forced Staycation

As for some individual travelers, the outlook in coming years could remain a bit more mixed and uncertain. In *The Future After COVID*, I wondered if people generally enjoyed having a forced staycation of sorts as a result of the COVID-19 "shelter in place" requirements — and if that could impact their decision to forego travel in the future, in exchange for more time at home.

After all, if they did enjoy the forced staycation, it may mean that in the future — based on the positive experience of having had a mandated staycation — some people may actually prefer them over vacations with travel to exotic places. This was not a guarantee or a hypothesis. It was just one potential future outcome. The notion is based on the fundamental truth that whenever people have a new experience, there is a chance that some might have liked the forced station. Of course, some hated it. Nevertheless, some people might now prefer staycations in the future.

No matter the experience, one thing does seem certain, though: People are likely to be more cognizant of social distance in the future — long after COVID-19 vaccinations have been widespread.

Dynamics Likely to Be True

For some travel topics, this is an area where the future simply cannot be as clear, and we need to consider various alternative futures. But there are a couple things that seem very likely to be true when it comes to travel and leisure.

First, travel and leisure are generally funded by disposable income. This is true for individuals, and it is also true at the corporate level. This means that both personal and business funds for travel and leisure are likely to fall if the economy remains under pressure. Corporate travel departments clamp down hard and fast when cash flow slows. This means that even if the conference business could come back as COVID-19 fears and risks fade, companies may be slow to spend the money for conferences without direct ROI.

It could be especially grim for the travel industry if people don't get their jobs back in those industries. But even if workers in travel, tourism, and leisure get their jobs back, travel companies, like airlines, could see their profit margins fall — even after travel ramps back up. After all, airlines make their best margins on last-minute business travelers, as well as on business class and first-class seats. This is why when demand for business travel falls, so does the profitability of airlines.

Business class seats on international flights are also usually extremely profitable as well, which is why airlines compete so much for those long-haul routes. As global travel dynamics begin to improve in the next two years, we could see significant competition among airlines, that undercuts some of that profitability.

In general, the recovery of the airline industry is likely to be very choppy. After all, airlines may find themselves at times with too much capacity — and at other times with not enough pilots.

The truth is that not only might it take months — or longer — before international travel is "normal" again after the COVID vaccine is widely available, but in the future, international flight travel bans may be much more quickly implemented than they were during the COVID-19 outbreak.

Downside Cruise Line Risks

Cruise lines have been some of the hardest hit industries during the COVID-19 pandemic. While most businesses in the space were forced to cease operations, companies that have tried to restart their businesses have struggled against both COVID risks as well as travel-shy customer. It doesn't help the industry that some of the initial COVID risks are underscored by the outbreak on the *Diamond Princess*, which was one of the first flashpoints.

Historically, whenever there have been various disease outbreaks, cruise lines tend to get caught in the middle of it. In general, cruises are the highest-value form of travel. After all, you get to go to numerous destinations, and the food is included.

All in all, the value of cruising is exceptionally high. This is one of the reasons it had become very popular in recent years. However, you do put people quite literally in closer quarters. And those close quarters become a greater risk in a pandemic-type scenario, which is what happened on the *Diamond Princess* at the beginning of the coronavirus outbreak.

Looking ahead to 2021 and potentially for several years beyond, we expect there might be significantly reduced demand for cruises. This drop in demand wouldn't just stem from a drop in disposable income, but it may also be a function of aggregate demand reduction at the population level, stemming from COVID-19 concerns and elevated social distancing norms.

Long-Term Professional Changes

There is another potential significant travel impact that could happen as a result of COVID-19 that has a multi-year behavioral impact. It seems likely that some heavy business travelers may opt to completely change their careers and business travel as a result of their experience following the outbreak of COVID-19.

In short, forced reductions of travel may lead some business travelers to change their careers, jobs, and lives to necessitate less travel. I know this has been the case with my own life in which I'm normally on the road almost nonstop. But as a result of the COVID-19 pandemic, I've been forced to change how I conduct business. If that were to be something I continue to do, it could drastically reduce my own travel. And if this occurs at the aggregate level, total overall travel demand may decline at the population level.

We may also see that in the same way that people increase their remote working and telecommuting as a result of the COVID-19 pandemic, they may also increase the amount of remote conferencing and remote meetings they do on a permanent basis after the vaccine, rather than doing as many in-person meetings as they had in the past.

And this isn't necessarily just a cost-reducing thing. It is also like the impact of commuting less. Because when you spend less time commuting, you spend less money on gasoline, and it saves you money. But more importantly, it saves you time — and that's especially true when we think about flying places for meetings.

After all, the greatest cost to a business traveler is not usually the plane ticket; it's their time. In truth, the hours of travel time both ways and the stress of business travel that comprises the greatest cost people might be trying to avoid. Now I know that you might be reading this and thinking, "I love travel so much."

But business travel is honestly a completely separate animal — and most businesspeople refer to business travel simply as *the road*. And let me tell you, the road is a grind. It was true before COVID and it will be still true long after COVID.

The Future of Travel and Leisure

We saw significant negative impacts to travel in 2020. And we are going to see some significant changes in the way people travel in the future. We will see people return to the skies — but it might be slower than people think.

For those that took to the road and camped, bought an RV, or discovered the joys of staycationing, their consumption habits may be changed for a long time to come. So, even though travel will likely increase sharply in the year ahead as COVID vaccines get deployed, we could still see a societal shift in the way people perceive travel in the long run.

And the way businesses consume travel could be reduced for some time in a way that could have negative impacts for travel companies overall. Because even if there are not outright travel declines ahead, the growth rates of travel beyond 2022 could be slower than they were before the COVID pandemic started.

The new, new normal of travel is likely to be a lot stronger than in 2020. The best way to gauge this recovery will be to watch the U.S. TSA throughput capacity as seen in Figure 4-5, but the comparison in the year ahead should be 2021 throughput compared to 2019 levels. I expect that even though 2021 is likely to be a year of improvements for travel and leisure, it will still be quite different than the world of travel as we knew it in 2019.

CHAPTER 17

THE FUTURE OF ESG AND SUSTAINABILITY

The push toward remote work during the COVID-19 pandemic has presented a unique opportunity for companies to claim ESG and sustainability progress. Looking ahead to the post-pandemic period, many companies will be unwilling to wind back that progress after the vaccine has been distributed.

One of the big trends for the future of finance is a push to incorporate environmental, social, and governance (ESG) goals along with sustainability. This is a trend that has been underway for some years and in the wake of the COVID-19 pandemic and shutdown, many companies have drastically reduced their carbon footprints. People are traveling less and commuting less, and there may be incentives for companies to show ESG progress that could keep many people working from home — and keep travel depressed for many years after COVID vaccinations.

In 2018, activist investors pushed for ESG goals and initiatives in climate change (19%), sustainability (13%), other environmental (7%), and political activity (19%).[1]

In other words, if we consider sustainability a type of environmental initiative along with climate change and other environmental 39 percent — a plurality — of all activist investor resolutions filed in 2018 were environment related.

Activist investors are usually large investors that use their shareholder power to push companies to make fundamental changes in the way they operate. Their activities have been on the rise. In fact, the number of companies subject to activist demands globally rose by almost 54 percent between 2013 and 2018. This can be seen in Figure 17-1. Domestically, the situation is very similar, and the number of U.S. companies subject to activist demands in 2018 was up by over 50 percent since 2013. This can be seen in Figure 17-2.

Activist investor demands fell in the United States and globally in 2019. But looking ahead beyond COVID, I expect the number of activist investor demands, as well as the number of companies impacted, are likely to continue rising on trend.

Externalities

Companies often benefit from things they don't pay for — but that still have costs. These costs that are not captured as business expenses but are instead passed on to society at large are called externalities. These include potentially negative environmental impacts of business operations. But they can also involve leveraging certain labor, societal, or political inefficiencies and imbalances to capture an arbitrage opportunity to reap outsized financial benefits. The data shows there is a push to make companies price-in these externalities; to make them pay.

Figure 17-1: Global Companies Subject to Activist Demands[2]

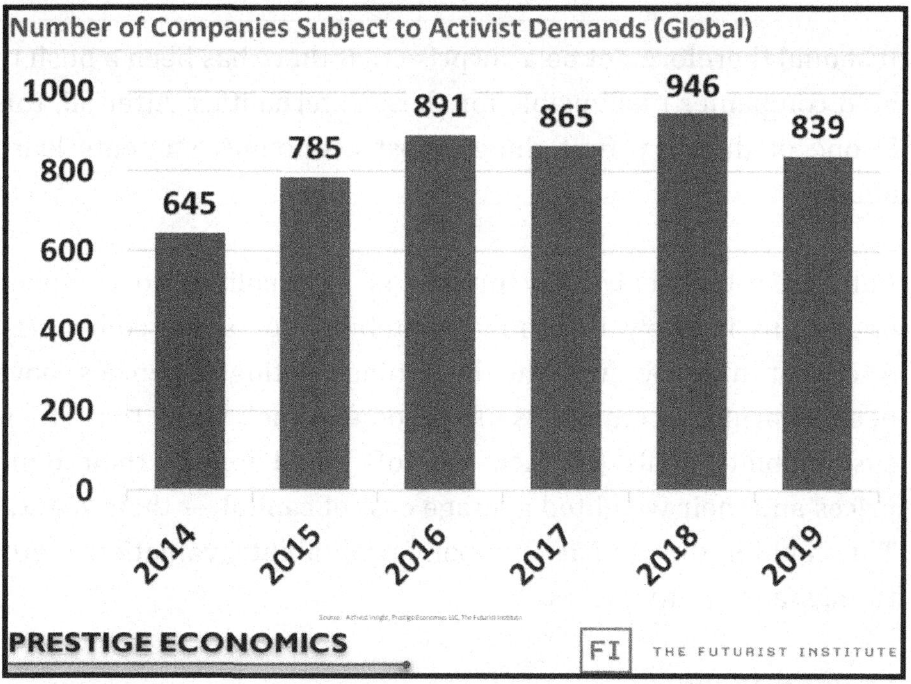

Figure 17-2: U.S. Companies Subject to Activist Demands[3]

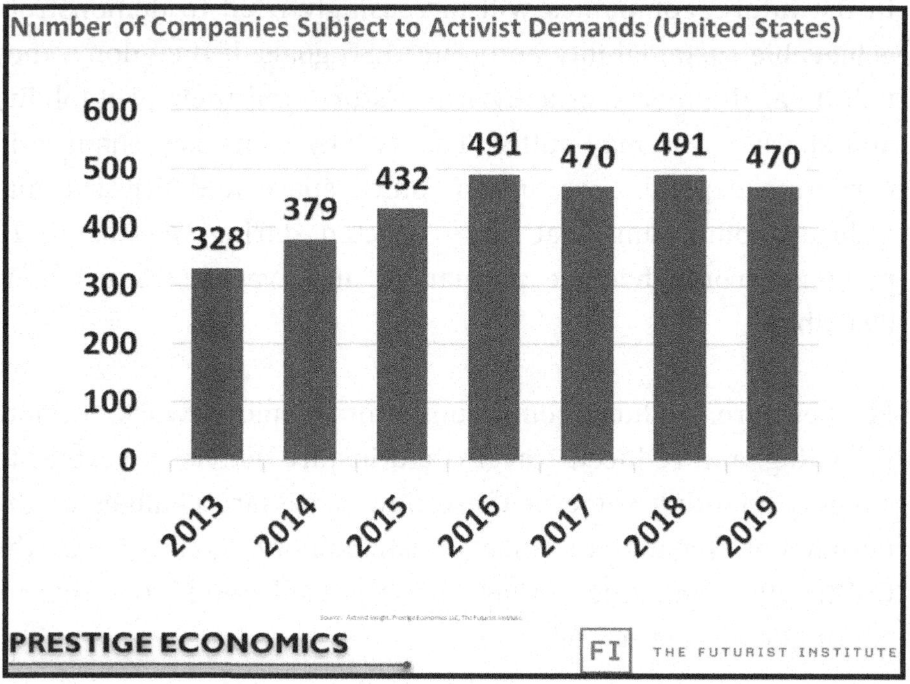

It should therefore not be a surprise that there has been a push to hold companies responsible for these externalities. After all, this is one of the very first things most economics students learn about.

But here's the deal: Fully pricing in externalities to company operations is likely to erode profitability for some companies. And that may be just the beginning. Ratings agencies have been warning companies for some time now that their sustainability goals (or lack thereof) could impact their bond prices and their weighted average cost of capital — their WACC.[4] This could in turn impact company profitability, valuation, credit ratings, and equity prices.

Looking Ahead

In the future, companies will increasingly need to demonstrate achievable sustainability and other ESG goals. If they don't, they will be at the mercy of activist investors, and their profitability and share prices may suffer. This is why some key changes in corporate expenditures, use of office space and utilities, and reduced commuting that were enacted during the COVID-19 pandemic could become permanent after vaccines are widely available.

Furthermore, reduced emissions from remote work, remote schooling, and reduced travel could inspire activist investors to support adopting some of these now necessary changes as the permanent means to meet sustainability goals once the COVID-19 pandemic crisis has passed and widespread vaccination has occurred.

The logic to making these changes permanent stems from a belief that if a company wants to show it is green, it might continue to implement remote work policies first enacted at scale during the COVID-19 pandemic, including:

— Reducing commuter fuel consumption of the office staff.
— Reducing power consumption in the office.
— Reducing water and paper product consumption in the office.

Outside the office, there could also be a push to reduce jet fuel consumption and the CO2 footprint that accompanies extensive (and especially international) business travel. The push to more sustainable flying requirements could reduce individual client visits, while simultaneously supporting demand for conferences, which are much more economical and efficient in terms of travel.

As in other areas, the COVID-19 pandemic has revealed for ESG and sustainability a potential that already existed but was not being acted on. In this case, the potential revealed was the potential to materially reduce energy consumption and emissions. Sustainability and climate change are likely to remain at the top of the shortlist for activity investor demands and despite the potential for companies to embrace more sustainability initiatives long after COVID vaccines have been widely distributed, I still expect the number of activist investor initiatives to increase on trend over time.

CHAPTER 18

THE FUTURE OF HEALTHCARE

COVID-19 shined a spotlight on healthcare. This was already a sector that has been growing in importance, especially given global demographic shifts toward lower birthrates and aging populations. This is true in the Unites States, Europe, Korea, Japan, and most of the planet.

As I discussed in Chapter 6, many young people who have lived through this pandemic may see it as a call to action and a reason to pursue a career in medicine, nursing, or care giving. This is a good thing, too, because there is a high — and rising — shortage of healthcare workers in the United States and globally.

In addition to the potential addition of new workers to the medical field, digital transformation has opened up a whole world of telemedicine and telehealth that is likely here to stay. Whether people are seeking treatments for physical or psychological issues, they will very likely continue to have an ability to access care givers remotely. In the same way that conferences, education, and work may be a new, new normal hybrid of access, the same is likely to be true of healthcare.

This could turn out to be especially beneficial for markets that have been inefficient. In other words, in areas where there are shortages of caregivers for physical or psychological treatments, the world is now the potential bullpen for those skilled caregivers. Plus, you no longer need to go into a doctor's office with one sickness, only to catch a staph infection or something else while you are there. Finally, the convenience of access could even reduce long-term medical care and insurance costs by making it easier for people to access care sooner — before needs become critical.

Looking ahead, there are likely to be significant policy changes in the field of healthcare in the United States under incoming President Biden. Millions of Americans didn't just lose their jobs in the pandemic; millions also lost their healthcare. This is something Biden spoke about as a presidential candidate, and it seems very likely that he will push for universal health care or Medicare for all. It would be almost impossible if the Republicans maintain control of the Senate. But that will be decided soon enough, on 5 January 2021. Even if Republicans hold the Senate after those runoffs, Democrats have a high chance of taking control of the Senate in the midterm elections of 2022, when more Republican seats are up for re-election than Democrat seats. Plus, many Republican elections will be in hotly contested battleground states. This means Republicans could lose whatever slim majority they are able to gain in the January 2021 runoffs.

No matter what happens politically in the United States, healthcare, health insurance, and access to care will remain in sharp focus in the U.S. and globally following the COVID-19 pandemic.

CONCLUSION

AFTER THE COVID VACCINE

My main goal in writing this book was to share my views of the potential future dynamics as a COVID-19 vaccine becomes broadly available. I have laid out some of the most important expectations, risks, and opportunities for the year ahead and beyond.

Of course, the COVID-19 vaccination situation is rapidly evolving, and many of the issues in this book may change drastically — or reach some level of resolution — more quickly than anticipated, depending on those developments. One thing seems certain, however: Many industries, businesses, individuals, and economies will be change in the year ahead as vaccines are distributed. There are opportunities for long-run positive impacts in the way we work, access education, prioritize the strength of our supply chain, improve access to food and paper products, and secure the future quality of (and access to) healthcare we experience. The future is uncertain, so it is critical to look for long-term trends with the greatest potential to persist in many different alternative futures scenarios. I hope this book helped you achieve those ends.

The Future is for Futurists

Futurists are going to be increasingly important for helping people and organizations think about the future. Of course, people define futurists in different ways. And there are various folks who consider themselves futurists, and they fall into a few camps. There are academic futurists who produce academic literature. There are people who call themselves futurists who are essentially future fanboys. Finally, there are applied futurists like me.

After all, the future is not certain. But we need to think about how our expectations of the future align with fundamentals of human nature, technological development, and historical trends. It's also important to frame discussions in terms of technologies that are likely to be broadly adopted over the coming decade — as well as those that are likely to be more important beyond that time window.

The team at The Futurist Institute and I refer to this time dichotomy as the "almost now" / "maybe someday" dynamic. Knowing what is going to be most critical in a near-term time window is just as critical as being able to identify those that could be more important farther in the future.

For example, expecting more broad-based remote work is something we have seen in the "almost now" 10-year window for the past four years. But remote working off-planet is something we continue to see as "maybe someday." In truth, limiting the number of "maybe someday" discussion points is critical for getting corporate and organization buy-in.

Long-term analysts and strategic planning around trends and technology are going to become more critical going forward. For this reason, as we look ahead at the future, we expect being a futurist will become an increasingly important profession.

Further Learning

If you've enjoyed this book and want to learn more about planning strategically for an uncertain future as well as alternative futures, I recommend pursuing the Certified Futurist and Long-Term Analyst™ — FLTA™ — training program that I created for The Futurist Institute.

The goal of this certification is to help analysts, executives, and professionals incorporate new and emerging trends and technologies into long-term strategic planning.

The FLTA™ certification program has six distinct professional tracks, including consulting, national security, financial planning, accounting, legal, and standard tracks. Plus, The Futurist Institute is accredited by the Certified Financial Planner Board of Standards® as a provider of continuing education hours. And the FLTA program includes 8.5 hours of CFP® continuing education. It is also eligible for continuing education hours from other organizations as well.

All of the details about the FLTA™ and The Futurist Institute can be found at www.futuristinstitute.org.

Your Next Steps

The end of the COVID-19 pandemic is not far ahead. Until vaccination reaches scale, however, the most important thing you can do is try and keep yourself and your loved ones out of harm's way. The importance of social distancing has been stressed repeatedly by healthcare professionals and public health policy leaders.

Recovery will come. This crisis, too, shall pass.

When it does, embracing the digital transformations of 2020 will be critical for success. This includes accepting that some of the short-term adaptations implemented during the time of the COVID-19 pandemic may be permanent — and some will not.

Preparing for second-order and third-order financial market and economic fallout will also be critical. If you look for opportunities to help your organization improve — or if you find ways to adapt and reshape your career — in the wake of this tragedy, you may be able to hasten the speed of that recovery.

You may even be able to find more upside in the long run.

Good luck and be well!

~ Jason Schenker
December 2020

FI

ENDNOTES

Chapter 2

1. Persico, Joseph E. (2005). *11th Month, 11th Day, 11th Hour* (illustrated, reprint ed.). London: Random House.

Chapter 4

1. November 2020 PMI data retrieved from the Institute of Supply Management, IHS Markit, TradingEconomics.com, and Econoday.com.
2. Ibid.
3. U.S. Census Bureau, E-Commerce Retail Sales as a Percent of Total Sales [ECOMPCTSA], retrieved from FRED, Federal Reserve Bank of St. Louis; https://fred.stlouisfed.org/series/ECOMPCTSA, 7 December 2020.
4. MHI BAI by Prestige Economics, November 2020.
5. TSA travel checkpoint numbers for 2019 and 2020. Retrieved on 15 December 2020 from https://www.tsa.gov/coronavirus/passenger-throughput.
6. Unemployment Insurance Weekly Claims. Retrieved on 18 December 2020 from https://www.dol.gov/ui/data.pdf.

Chapter 6

1. Bureau of Labor Statistics. Retrieved on 18 December 2020 from https://www.bls.gov/cps effects-of-the-coronavirus-covid-19-pandemic.htm#table1.
2. Ibid.
3. "2017 State of Telecommuting in the U.S. Employee Workforce." Flexjobs. Retrieved on 9 May 2019 from https://www.flexjobs.com/2017-State-of-Telecommuting-US.
4. Ibid.
5. Unemployment rate for persons 25 years and older by educational attainment. Bureau of Labor Statistics. Retrieved on 15 December 2020 from https://www.bls.gov/charts/employment-situation/unemployment-rates-for-persons-25-years-and-older-by-educational-attainment.htm.
6. Bureau of Labor Statistics. Retrieved on 15 December 2020 from https://www.bls.gov/opub/ted/2016/24-percent-of-employed-people-did-some-or-all-of-their-work-at-home-in-2015.htm.
7. U.S. Bureau of Labor Statistics, Unemployment Rate - College Graduates - Master's Degree, 25 years and over [CGMD25O], retrieved from FRED, Federal Reserve Bank of St. Louis; https://fred.stlouisfed.org/series/CGMD25O, 15 December 2020.
8. Bureau of Labor Statistics. Retrieved on 15 December 2020 from https://www.bls.gov/ooh/most-new-jobs.htm.
9. Bureau of Labor Statistics: Retrieved on 15 December 2020 from https://www.bls.gov/ooh/fastest-growing.htm.

Chapter 7

1. U.S. Bureau of Labor Statistics, Consumer Price Index for All Urban Consumers: All Items in U.S. City Average [CPIAUCSL], retrieved from FRED, Federal Reserve Bank of St. Louis; https://fred.stlouisfed.org/series/CPIAUCSL, 14 December 2020.
U.S. Bureau of Labor Statistics, Consumer Price Index for All Urban Consumers: Medical Care in U.S. City Average [CPIMEDSL], retrieved from FRED, Federal Reserve Bank of St. Louis; https://fred.stlouisfed.org/series/CPIMEDSL, 14 December 2020.
2. Ibid.
3. Department of Education. Retrieved on 14 December 2020 from https://nces.ed.gov/programs/digest/d18/tables/dt18_206.10.asp.
4. Ibid.
5. Bureau of Labor Statistics. Retrieved on 14 December 2020 from https://www.bls.gov/emp/graphics/2019/unemployment-rates-and-earnings.htm.

Chapter 9
1. Board of Governors of the Federal Reserve System (US), Assets: Total Assets: Total Assets (Less Eliminations from Consolidation): Wednesday Level [WALCL], retrieved from FRED, Federal Reserve Bank of St. Louis; https://fred.stlouisfed.org/series/WALCL, 14 December 2020.
2. Federal Reserve. "The Federal Reserve's Monetary Policy Toolkit: Past, Present, and Future." Retrieved from https://www.federalreserve.gov/newsevents/speech/yellen20160826a.htm.
3. Board of Governors of the Federal Reserve System (US), Effective Federal Funds Rate [FEDFUNDS], retrieved from FRED, Federal Reserve Bank of St. Louis; https://fred.stlouisfed.org/series/FEDFUNDS, 14 December 2020.
4. "Genesis Block." Wikipedia. Bitcoin Wiki. Retrieved on 24 August 2018 from https://en.bitcoin.it/wiki/Main_Page.

Chapter 10
1. U.S. Department of the Treasury. Fiscal Service, Federal Debt: Total Public Debt [GFDEBTN], retrieved from FRED, Federal Reserve Bank of St. Louis; https://fred.stlouisfed.org/series/GFDEBTN, 14 December 2020.
2. Ibid.
3. Federal Reserve Bank of St. Louis and U.S. Office of Management and Budget, Federal Debt: Total Public Debt as Percent of Gross Domestic Product [GFDEGDQ188S], retrieved from FRED, Federal Reserve Bank of St. Louis; https://fred.stlouisfed.org/series/GFDEGDQ188S, July 21, 2020.
4. Gopinath, G. (24 June 2020). IMF Blog: *Reopening from the Great Lockdown: Uneven and Uncertain Recovery.* International Monetary Fund. Retrieved on 20 July 2020 from https://blogs.imf.org/2020/06/24/reopening-from-the-great-lockdown-uneven-and-uncertain-recovery.
5. Mayer, J. (18 November 2015). "The Social Security Façade." Retrieved 24 July 2020: https://www.usnews.com/opinion/economic-intelligence/2015/11/18/social-security-and-medicare-have-morphed-into-unsustainable-entitlements.
6. U.S. Social Security Administration. (September 2017). *Fast Facts and Figures About Social Security, 2017,* p. 8. Retrieved on 21 July 2020: https://www.ssa.gov/policy/docs/chartbooks/fast_facts/2017/.
7. U.S. Social Security Administration. (July 2020). *Fast Facts and Figures About Social Security, 2020,* p. 37. Retrieved on 21 July 2020: https://www.ssa.gov/policy/docs/chartbooks/fast_facts/2020/fast_facts20.pdf.
8. Ibid.
9. Ibid.
10. Ibid.
11. World Bank, Population Growth for the United States [SPPOPGROWUSA], retrieved from FRED, Federal Reserve Bank of St. Louis; https://fred.stlouisfed.org/series/SPPOPGROWUSA, June 5, 2018.
12. Last, J. (2013) *What to Expect, When No One's Expecting: America's Coming Demographic Disaster.* New York: Encounter Books, pp. 2-4.
13. Ibid., p. 3.
14. Ibid., p. 109.
15. 16. U.S. Social Security Administration. Retrieved 11 February 2017 from https://www.ssa.gov/history/ratios.html.
16. Coy, Peter. (29 July 2020). "Americans Aren't Making Babies and That's Bad for the Economy." Bloomberg. Retrieved on 15 August 2020 from https://www.bloomberg.com/news/articles/2020-07-29/coronavirus-pandemic-americans-aren-t-making-babies-in-crisis.

Chapter 11

1. "Listed Domestic Companies, Total." *Data*, World Bank, data.worldbank.org/indicator/CM.MKT.LDOM.NO?locations=US.
2. Ritter, Jay R. "IPO Data." *Warrington College of Business*, University of Florida. Retrieved on 15 August 2020 from site.warrington.ufl.edu/ritter/ipo-data/.
3. Ibid.
4. Ibid.
5. Ibid.
6. Ibid.

Chapter 12

1. Markowitz, A. (23 November 2020). "13 Iconic Retailers that have Fallen into Bankruptcy." Retrieved on 13 December 2020 from https://www.aarp.org/money/credit-loans-debt/info-2020/bankrupt-retail-chain-store-list-is-growing.html

Chapter 17

1. Welsh, H. (9 November 2018). "Social, Environmental & Sustainable Governance Shareholder Proposals in 2018." *Securities and Exchange Commission*, Sustainable Investments Institute. Retrieved on 12 July 2019 from www.sec.gov/comments/4-725/4725-4636528-176443.pdf.
2. "The Activist Investing Annual Review." (2020). *Activist Insight.* Retrieved on 12 August 2020 from https://www.activistinsight.com/research/TheActivistInvesting_AnnualReview_2020.pdf.
3. Ibid.
4. This comes on the heels of pressure from activist investor pushes on the Exxon board for a lack of ESG oversight. "Exxon Board Targeted for Lack of ESG Oversight." (May 11, 2019). National Association of Corporate Directors. Retrieved on 12 July 2019 https://tinyurl.com/NACDExxon2019.

ABOUT THE AUTHOR

Mr. Schenker is the President of Prestige Economics and Chairman of The Futurist Institute. He has been ranked one of the most accurate financial forecasters and futurists in the world. Bloomberg News has ranked Mr. Schenker a top forecaster in 44 categories, including #1 in the world for his accuracy in 26 categories, including for his forecasts of the Euro, the British Pound, the Russian Ruble, the Chinese RMB, crude oil prices, natural gas prices, gold prices, industrial metals prices, agricultural commodity prices, and U.S. jobs.

Mr. Schenker's work has been featured in *The Wall Street Journal*, *The New York Times*, and the *Frankfurter Allgemeine Zeitung*. He has appeared on CNBC, CNN, ABC, NBC, MSNBC, Fox, Fox Business, BNN, Bloomberg Germany, and the BBC. Mr. Schenker has also been a guest host of Bloomberg Television, and he is a contributor for *Bloomberg Opinion*. Mr. Schenker was ranked one of the top 100 most influential financial advisors in the world by Investopedia in 2018.

Mr. Schenker attends OPEC and Fed events, and he has given keynotes for private companies, public corporations, industry groups, and the U.S. Federal Reserve. He has advised NATO and the U.S. government on the future of work, blockchain, Bitcoin, cryptocurrency, quantum computing, data analysis, forecasting, and fake news. Mr. Schenker has written 32 books. Twelve have been bestsellers, including: *The Future After COVID, Jobs for Robots, Quantum: Computing Nouveau, Commodity Prices 101, Recession-Proof, Futureproof Supply Chain, Electing Recession, The Future of Finance is Now, The Future of Energy, The Dumpster Fire Election,* and *The Robot and Automation Almanac* for 2018 and 2020. Mr. Schenker also wrote *The Promise of Blockchain, Futureproof Supply Chain, The Future of Food and Agriculture, The Fog of Data, Financial Risk Management Fundamentals, Midterm Economics, Spikes: Growth Hacking Leadership, Strategic Cost-Cutting, Strategic Cost-Cutting After COVID,* and *Reading the Economic Tea Leaves.* Mr. Schenker was featured as one of the world's foremost futurists in *After Shock.* His futurist research was included in an unclassified Pentagon report in June 2020.

Mr. Schenker advises executives, industry groups, institutional investors, and central banks as the President of Prestige Economics. He also founded The Futurist Institute, for which he created a rigorous course of study that includes The Future of Work, The Future of Transportation, The Future of Data, The Future of Finance, Futurist Fundamentals, The Future of Energy, The Future of Leadership, The Future of Healthcare, The Future of Quantum Computing, and The Future After COVID. Mr. Schenker is also an instructor for LinkedIn Learning courses on Economic Indicators, Risk Management, Audit and Due Diligence, and Recession-Proof Strategies.

Mr. Schenker holds a Master's in Applied Economics from UNC Greensboro, a Master's in Negotiation, Conflict Resolution, and Peacebuilding from CSU Dominguez Hills, a Master's in Germanic Languages and Literature from UNC Chapel Hill, and a Bachelor's in History and German from The University of Virginia. He also holds a Certificate in FinTech from MIT, a Certificate in Supply Chain Management from MIT, a Certificate in Professional Development from UNC, a Certificate in Negotiation from Harvard Law School, a Certificate in Cybersecurity from Carnegie Mellon, and a Professional Certificate in Strategic Foresight from the University of Houston. Mr. Schenker holds the designations CMT® (Chartered Market Technician), and CFP® (Certified Financial Planner). He is also a Certified Futurist and Long-Term Analyst™ and holds the FLTA™ designation.

Before founding Prestige Economics, Mr. Schenker worked as a Risk Specialist at McKinsey and Company, where he provided content direction to trading, risk, and commodity project teams on six continents. Prior to McKinsey, Mr. Schenker was the Chief Energy and Commodity Economist at Wachovia, which is now Wells Fargo.

Based in Austin, Texas, Mr. Schenker is one of only 100 CEOs on the Texas Business Leadership Council, a non-partisan organization that advises Texas elected leadership at the state and federal level. Mr. Schenker is a Governance Fellow of the National Association of Corporate Directors. He also sits on multiple boards and is the VP of Finance on the Executive Committee of The Texas Lyceum, the preeminent non-partisan leadership group in Texas. Mr. Schenker is also a member of the Bretton Woods Committee.

THE FUTURIST INSTITUTE

FI THE FUTURIST INSTITUTE

The Futurist Institute was founded in 2016 to help analysts, executives, and professionals incorporate technology and trend risks and opportunities into strategic planning. The Futurist Institute confers the Futurist and Long-Term Analyst™ (FLTA) designation and helps analysts become Certified Futurists™. Our courses have been approved for continuing education hours by the Certified Financial Planner Board of Standards (CFP Board).

Current Courses

The Future After COVID
The Future of Work
The Future of Data
The Future of Energy
The Future of Finance
The Future of Healthcare
The Future of Leadership
The Future of Transportation
Futurist Fundamentals
Quantum Computing
Write the Future

Visit The Futurist Institute:

www.futuristinstitute.org

PUBLISHER

Prestige Professional Publishing was founded in 2009 to produce insightful and timely professional reference books. We are registered with the Library of Congress.

Published Titles

After the COVID Vaccine
Be the Shredder, Not the Shred
Commodity Prices 101
Electing Recession
Financial Risk Management Fundamentals
Futureproof Supply Chain
A Gentle Introduction to Audit and Due Diligence
Jobs for Robots
Jobs for Robots: COVID Edition
Midterm Economics
Quantum: Computing Nouveau
Reading the Economic Tea Leaves
Recession-Proof Career Strategies After COVID
Robot-Proof Yourself
Spikes: Growth Hacking Leadership
Strategic Cost-Cutting
Strategic Cost-Cutting After COVID
The Dumpster Fire Election
The Economics of Revolt and Revolution
The Fog of Data
The Future After COVID

PUBLISHER

Published Titles

The Future of Agriculture
The Future of Energy
The Future of Finance is Now
The Future of Food and Agriculture After COVID
The Promise of Blockchain
Write the Future
The Robot and Automation Almanac - 2018
The Robot and Automation Almanac - 2019
The Robot and Automation Almanac - 2020
The Robot and Automation Almanac - 2021

Forthcoming Titles

Content Monster
Disruption Warfare
The Future of Healthcare
The Future of Marketing
The Future of Travel and Leisure

THE FUTURE AFTER COVID

In *The Future After COVID*, top-ranked forecaster and Chairman of The Futurist Institute Jason Schenker offers a futurist perspective into the potential long-term changes, challenges, and opportunities that COVID-19 is likely to have for over a dozen different critical fields and industries. The book was published in April 2020. It has been an international bestseller.

JOBS FOR ROBOTS

Jobs for Robots: COVID Edition provides an in-depth look at the future of automation and robots, with a focus on opportunities and risks for the future following the COVID-19 pandemic. The first edition of *Jobs for Robots* was published in February 2017. It has been a #1 Best Seller on Amazon.

FROM THE AUTHOR

The following disclaimer applies to any content in this book:

This book is commentary intended for general information use only and is not investment advice. Jason Schenker does not make recommendations on any specific or general investments, investment types, asset classes, non-regulated markets, specific equities, bonds, or other investment vehicles. Jason Schenker does not guarantee the completeness or accuracy of analyses and statements in this book, nor does Jason Schenker assume any liability for any losses that may result from the reliance by any person or entity on this information. Opinions, forecasts, and information are subject to change without notice. This book does not represent a solicitation or offer of financial or advisory services or products; this book is only market commentary intended and written for general information use only. This book does not constitute investment advice. All links were correct and active at the time this book was published.

DISCLAIMER

FROM THE PUBLISHER

The following disclaimer applies to any content in this book:

This book is commentary intended for general information use only and is not investment advice. Prestige Professional Publishing, LLC does not make recommendations on any specific or general investments, investment types, asset classes, non-regulated markets, specific equities, bonds, or other investment vehicles. Prestige Professional Publishing, LLC does not guarantee the completeness or accuracy of analyses and statements in this book, nor does Prestige Professional Publishing, LLC assume any liability for any losses that may result from the reliance by any person or entity on this information. Opinions, forecasts, and information are subject to change without notice. This book does not represent a solicitation or offer of financial or advisory services or products; this book is only market commentary intended and written for general information use only. This book does not constitute investment advice. All links were correct and active at the time this book was published.

Prestige Professional Publishing, LLC

4412 City Park Road #4

Austin, Texas 78730

www.prestigeprofessionalpublishing.com

PRESTIGE
PROFESSIONAL PUBLISHING

ISBN: 978-1-946197-96-2 *Paperback*

978-1-946197-72-6 *Ebook*

Made in United States
Orlando, FL
10 November 2026

90498871R00105